Victor Vazquez
Jose R. Parga
Jesus L. Valenzuela

Cyanide and Arsenic Destruction by Electrocoagulation

Victor Vazquez
Jose R. Parga
Jesus L. Valenzuela

Cyanide and Arsenic Destruction by Electrocoagulation

Theory of Cyanide Destruction with Anatase and Arsenic Adsorption on Magnetic Electrocoagulation Species

ScienciaScripts

Cover image: www.ingimage.com

This book is a translation from the original published under ISBN 978-613-9-05327-8.

Publisher:
Sciencia Scripts
is a trademark of
Dodo Books Indian Ocean Ltd. and OmniScriptum S.R.L publishing group

120 High Road, East Finchley, London, N2 9ED, United Kingdom
Str. Armeneasca 28/1, office 1, Chisinau MD-2012, Republic of Moldova, Europe
Managing Directors: Ieva Konstantinova, Victoria Ursu
info@omniscriptum.com

Printed at: see last page
ISBN: 978-620-8-36678-0

Contents

SUMMARY

The adsorption of titanium dioxide and arsenic on electrocoagulation generated species such as magnetite, goethite and lepidocrocite was studied.

Titanium dioxide was used to remove cyanide from water (400 ppm CN^-) using a photocatalytic process. A 94 % cyanide removal rate was achieved with this process. However, this technique has a disadvantage for industrial use, the separation of titanium dioxide after cyanide removal is difficult due to the fineness of the particles, therefore, the reuse of titanium dioxide has not been advanced for the treatment of cyanide contaminated water. To remedy this point, the Electrocoagulation (EC) process was used to recover the titanium dioxide from the solution. Using EC a 99 % recovery rate was achieved.

On the other hand, mining and metallurgical production has generated waste containing heavy metals and their by-products. Specifically, arsenic compounds have been discharged into the environment. EC was used as a viable alternative to treat this pollutant, with this process 99 % arsenic removal was obtained without the addition of any other chemical reagents.

The Langmuir isotherm and the Langmuir-Hinshelwood equation were used to determine the feasibility of adsorption of titanium dioxide and arsenic on EC generated species. Thermodynamic and kinetic parameters such as free energy (AG^o), enthalpy (AH^o), entropy (AS^o), adsorption constant and kinetics were calculated.

OBJECTIVES

General Objective

To carry out a kinetic and thermodynamic study of the adsorption of titanium dioxide and arsenic on species generated by electrocoagulation such as magnetite, goethite, lepidocrocite.

Particular Objectives

Currently, precious metal recovery processes inevitably require the use of cyanide solutions. As the leaching process proceeds, the concentration of cyanide decreases to a point where the concentration is such that the solution ends its useful life in the process and must be discarded, creating serious environmental problems. In this study we propose the use of an innovative technique which will help us to have less toxic waste. In order to achieve this, the present research work proposes:

1. Destroy cyanide in an effluent containing this pollutant using titanium dioxide nanocrystals by photocatalytic oxidation and subsequently recover the titanium dioxide using the electrocoagulation process.
2. This study will also analyse the removal of arsenic from pollutant effluents using the electrocoagulation process and finally a kinetic and thermodynamic study will be carried out to determine the mechanisms of the adsorption of nanometric titanium dioxide particles and arsenic ions on the magnetite and goethite species generated on the iron electrodes during the electrocoagulation process.
3. This analysis will extend the knowledge of the electrocoagulation process and specifically, the adsorption of titanium dioxide and arsenic particles on iron species. For this theoretical analysis, the Langmuir-Hinshelwood equation will be used to characterise the mechanism.

I INTRODUCTION

Currently, Mexico is the world's leading producer of silver and according to statistics it has an annual production of 2,103,125 tons[1] . Our country, like the rest of the world, uses the traditional cyanidation method to recover gold and silver, in the process cyanide effluents are generated which are dangerous for the environment. The chemical reaction for the dissolution of gold and silver can be expressed by the Elsner equation :[2]

$$4Ag + 8CN^- + O_2 + 2H_2O \rightarrow 4[Ag(CN^-)_2] + 4OH^- \qquad (1.1)$$

After the extraction and recovery of gold and silver, large quantities of cyanide are disposed of in effluents creating environmental problems due to the hazardous nature of cyanide.[3].

In the United States the EPA has set a maximum effluent limit of 0.2 mg/litre for drinking water, in Germany and Switzerland the limit is 0.01 mg/litre and in Mexico the SEMARNAT has set the cyanide limit at 0.2 mg/litre[4] . Due to these considerations the recovery or destruction of cyanide is a necessary processing step. To reduce the levels of cyanide sent to effluents, several processes have been developed in their treatment. All of these methods are based on the recovery of cyanide by acidification or destruction by chemical oxidation, these processes have several disadvantages, for example the oxidation reagents are expensive and patent fees have to be paid .[5]

Due to these factors, the technique of photocatalytic oxidation with titanium dioxide is one of the innovative ways for the treatment of cyanide-contaminated water. However, this technique has a disadvantage for its industrial application: the separation of titanium dioxide nanoparticles after photocatalytic degradation of cyanide is difficult due to the fineness of the titanium particles and therefore, not much progress has been made in the reuse of titanium dioxide for the treatment of cyanide-contaminated water. To remedy this point, the electrocoagulation technique was used to recover the titanium dioxide from the aqueous suspension.

Another important problem is the contamination of soils and aquifers by arsenic, as is the case in Torreon Coahuila and other parts of Mexico[6] , as a result of metallurgical and industrial production. A study carried out at the University of Cincinnati reveals the existence of arsenic, cadmium and lead in the soil of areas near lead and zinc smelters, in concentrations that exceed national and international permitted levels. Because of the above, in this study we will use the electrocoagulation process to remove arsenic from contaminated water.

Although the EC process is technologically known, to date, nowhere in the world has it been characterised and analysed kinetically and thermodynamically, nor is it applied on an industrial scale for the recovery of TiO2, since the fundamental mechanisms and principles on which this method is based, both in the recovery of TiO2 particles and in the removal of arsenic from contaminated water, are still completely unknown. Therefore, a fundamental study is needed to determine the kinetic and thermodynamic mechanism of the adsorption of TiO2 nanoparticles and arsenic ions on the magnetite and goethite species, etc. that are generated on the iron electrodes in the EC reactor. For this purpose, the Langmuir-Hinshelwood equation will be used to characterise the mechanism.

II STATE OF THE FIELD OR STATE OF THE ART

For the development of this topic it is necessary first to analyse the chemical and physical principles of water pollutants. Afterwards, it is necessary to see which techniques have been used successfully and after the analysis, to propose a new technique with its respective description.

2.1 Description of the contaminant (cyanide)

Cyanide (CN^-) is an anion containing carbon and nitrogen bonded by a triple bond, is able to react easily, even at very low concentrations, with heavy metals, is a highly toxic substance that can be easily absorbed by tissues and is one of the most regulated hazardous substances in discharges to the environment as it is considered a Class - P hazardous waste substance by RCRA (Resource Conservation and Recovery Act)[7].

The main forms of cyanide are hydrogen cyanide (HCN), sodium cyanide (NaCN) and potassium cyanide (KCN). Cyanide can be found as a colourless gas (HCN and ClCN) or in the form of crystals (NaCN and KCN).

Cyanide compounds in which CN can be obtained as CN^- are classified as simple cyanides and complex cyanides. The simple ones are represented by the formula $A(CN)_x$, where A is an alkali element (sodium, potassium, ammonium) or a metal, and x, the valence of A, is the number of CN groups[8] . In aqueous solutions of simple alkaline cyanides, the CN group is present as CN^- and molecular HCN, the ratio depending on the pH and the dissociation constant for molecular HCN (pka ~9.2) (Figure 2.1). In most natural waters, HCN predominates. In solutions of simple metal cyanides the CN group can also appear in the form of anionic metal cyanide complexes of varying stability. Most of these simple complexes are poorly soluble or nearly insoluble (CuCN, AgCN, $Zn(CN)_2$), but these, in the presence of alkali cyanides, form a variety of highly soluble metal-cyanide complexes[8] . Alkali metal cyanides can usually be represented by $A_yM(CN)_x$. A represents the alkali element present, y the number of times the alkali element occurs, M the heavy metal and x the number of CN groups. The initial dissociation of each of these soluble cyanide complexes produces an anion which is the $M(CN)_{xy}$ radical. This can further dissociate depending on many factors, with the release of CN^- and the consequent formation of HCN .[8]

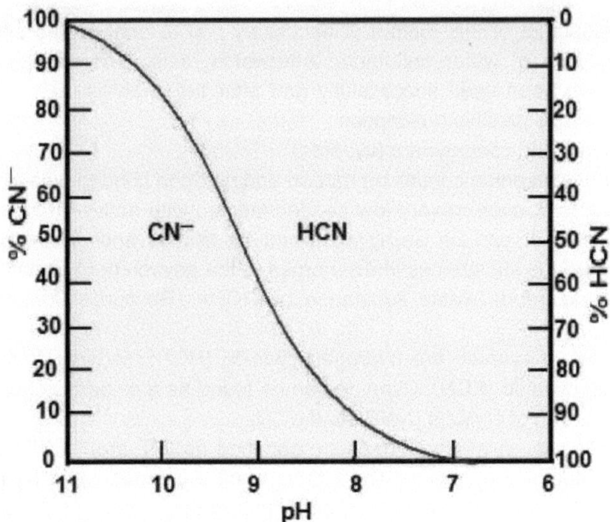

Figure 2.1. pH equilibrium between CN^- /HCN .[8]

2.2 Toxicity

The presence of free cyanides and cyanide complexes in industrial wastewater streams is a major problem due to the known toxicity of these species to living organisms, even at very low concentrations[9] . The toxicity of free cyanide is lower than that of HCN, which is formed by the hydrolysis of cyanide and is a potential risk to aquatic life because most of the water bodies where waste streams are disposed have a pH lower than the pKa of molecular HCN (<9.2), causing the cyanide in most of the discharges to be converted into its more toxic form .[10]

In all prokaryotic or eukaryotic cells (bacteria, fungi, plants, animals, including humans) a vital function is respiration. One of the indispensable molecules for this function is cytochrome-C oxidase, which has an iron (Fe) atom at the centre of its complex structure. When cyanide enters the cells it "captures" the Fe and the enzyme is no longer functional. The consequence is that the cell stops "breathing" and dies .[11]

Exposure to 300 ppm hydrogen cyanide can be fatal within minutes. Ingestion or absorption through wounds of 100 to 250 mg NaCN or KCN can also be fatal[10] . Cyanide is described as having a "bitter almond" odour, but it does not always emanate an odour and not all people can detect it[11] . In large quantities cyanide is very harmful to people. Exposure to high levels of cyanide in the air for a short period of time can result in brain damage, heart damage, coma and even death. Exposure to low levels of cyanide for long periods of time can result in breathing difficulties, heart attacks, vomiting, convulsions, blood changes, headaches and enlargement of the thyroid gland .[11]

2.3 The cyanidation process

The traditional method of recovering silver and gold is cyanidation (Figure 2.2), which generates cyanide effluents that are hazardous to the environment.

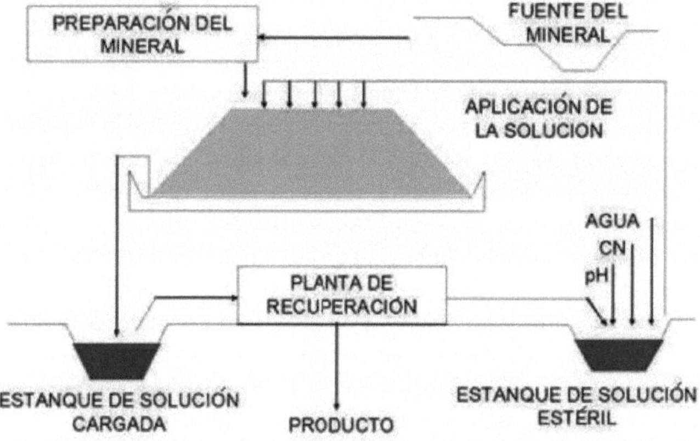

CHARGED SOLUTION PONDPRODUCT

Figure 2.2 Flow diagram of the heap cyanidation process .[12]

In the extractive industry, this method of precious metal recovery is carried out using a cyanide solution of 0.03-0.3 % NaCN with a pH above 10 to avoid the formation of HCN and also needs efficient aeration to have an oxygen concentration of more than 7 mg/litre in the pulp. The chemical reaction for the dissolution of gold and silver using the above conditions can be expressed by the Elsner equation:

$$4Au + 8CN^- + O_2 + 4H_2O = 4[Au(CN^-)_2] + 4OH^- \qquad (2.1)$$

This is carried out by the following mechanism:

$$Au^+ + 2\,CN^- = Au(CN)_2^- + e^- \qquad (2.2)$$

$$O_2 + 2\,H_2O + 2\,e^- = 2\,OH^- + H_2O_2 \qquad (2.3)$$

$$H_2O_2 + 2\,e^- = 2\,OH^- \qquad (2.4)$$

In this electrochemical mechanism, the cyanide ion complexes with the gold, and the oxygen acts as an oxidant (2.1). However, the association of silver with the cyanide ion is weaker than that of gold and the dissolution of silver requires longer contact time .[13]

After the extraction and recovery of gold and silver, large quantities of cyanide are disposed of in effluents creating environmental problems due to the hazardous nature of cyanide. In industrial effluents cyanide exists in three forms: free cyanide as HCN; simple cyanide as NaCN; complexed cyanide as $Fe(CN)_6^{-3}$, $Ni(CN)_4^{-2}$, $Zn(CN)_4^{-2}$ y $Cu(CN)_4^{-2}$. Total cyanide is the sum of free, simple and complexed cyanide where other ligands such as cyanate and thiocyanate are not taken into account .[13]

Due to the large number of industries that use cyanide in their processes, especially the mining and metallurgical industry that uses large quantities of this reagent, it is of utmost importance to establish treatment methods for the recovery or elimination of cyanide in the effluents in order to operate the process economically and safely.

In the United States the EPA has set the maximum effluent limit of 0.2 mg/litre, in Germany and Switzerland the limit is 0.01 mg/litre and in Mexico the SEMARNAT has set the cyanide limit at 0.2 mg/litre[14] . Due to these considerations, it is of utmost importance to develop more efficient technology for cyanide recovery or definitely cyanide removal depending on the total concentration of cyanide in the effluent.

There are two ways of dealing with these discharges to the environment, either by traditional separation methods or by remediation methods. Traditional separation methods basically transfer the toxic substance from one stream to another, so that the underlying problem persists. Remediation methods, on the other hand, seek to degrade these substances and transform them into non-toxic or less toxic substances. Many cyanide remediation processes are currently available, most of them do not treat all cyanide derivatives or complexes formed by cyanide, are very expensive or require further treatment.

2.4 Conventional oxidation methods for cyanide elimination

Conventional oxidation processes generally use reagents such as chlorine, ozone, hypochlorites, permanganates, peroxide, and some combinations of these. In these processes the organic pollutant is transformed by the oxidising action of these compounds, sometimes to harmless products such as CO_2 and H_2O. These methods are generally expensive because of the demand for reagents and in some cases patent rights have to be paid for[15] , their subsequent separation from the water and the control of the process requires particular care in the handling of the reagents.

Of the above mentioned methods, none achieves optimal removal, i.e. high levels of effluent purity with low consumption of chemical inputs and/or energy, therefore technological research worldwide in recent years has proposed detoxification by advanced oxidation processes as an efficient alternative .[16]

2.4.1 Alkaline chlorination

Alkaline chlorination is one of the most widely used methods for cyanide destruction. It is mainly used in the areas of metal deposition and in the steel industry. The principle of alkaline chlorination is the oxidation of cyanide to cyanate; the reagents used are chlorine gas (Cl_2), sodium hypochlorite (NaClO) or calcium hypochlorite ($Ca(ClO)_2$). The oxidation of cyanide with chlorine is carried out by the following reactions :[17]

Stage 1:

$$CN^- + Cl_2 = CNCl + Cl^- \quad (2.5)$$

$$CNCl + 2NaOH = NaCNO + H_2O + NaCl \quad (2.6)$$

Stage 2:

$$2NaCNO + 4NaOH + 3Cl_2 = 2CO_2 + N_2 + 6NaCl + 2H_2O \quad (2.7)$$

In general, alkaline chlorination is a powerful destruction method, even neutralising cyanometallic complexes (with the exception of ferrocyanides). Its main disadvantage is that the chlorine compounds are toxic and the reagents used are expensive.

2.4.2 Ozonisation

The ozone (O_3) required for this oxidation process is generated electrically from air or oxygen. The use of oxygen has the advantage of providing twice the concentration of ozone over the use of air, with half the power requirement of air, and provides an additional contribution to the oxidation of cyanide due to the higher concentration of oxygen.

The reactions in this process are as follows :[17]

$$CN^- + O_3 = CNO^- + O_2 \quad \text{(rápida)} \qquad (2.8)$$

$$CNO^- + O_3 + 2H_2O = NH_3 + HCO_3^- + 1.5O_2 \quad \text{(lenta)} \qquad (2.9)$$

Its main disadvantage is the cost of the equipment to produce ozone.

2.4.3 INCO (sulphur dioxide)

The INCO (International Nickel Company) process for the removal of cyanide from industrial waste effluents was patented by G.J. Borboly in 1984 .[17]

The process uses gaseous SO2 (or solutions of sodium sulphite, Na2SO3 or Na2S2O5) in combination with air as oxidant and lime for pH control. Copper (as Cu^{+2}) in the solution is used as a catalyst for the oxidation of the cyanide. This process has the widest industrial application. Cyanide oxidation is carried out with the following overall reaction :[17]

$$CN^- + SO_2 + O_2 + H_2O = CNO^- + H_2SO_4 \qquad (2.10)$$

The SO2 requirement is 2.47 g SO2 per gram of CN^- , the optimum pH levels are between 9 and 10. The main disadvantages of this process are the patent fees, the addition of sulphates, and the proper disposal of ferrocyanides, if they precipitate.

2.4.4 Oxidation with hydrogen peroxide (H2O2)

Hydrogen peroxide is an oxidising agent used in the treatment of organic effluents, sulphides and cyanides. There are two processes that employ peroxide for the destruction of cyanide, both of which work under alkaline conditions (8<pH<=12) [17]).:

1) With copper catalyst:

$$CN^- + H_2O_2 \xrightarrow{Cu^{++}} CNO^- + H_2O \qquad (2.11)$$

$$CNO^- + 2H_2O \xrightarrow{pH<7} NH_4^+ + CO_3^{-2} \qquad (2.12)$$

2) Dupont-Keastone process:

$$HCN + HCHO \xrightarrow{pH-11, Cu} HOCH_2CN \qquad (2.13)$$

(glicolonitrilo)

$$HOCH_2CN + H_2O_2 \xrightarrow{pH-11} 3NH_35HCNO5H_2C(OH)CONH_2 \qquad (2.14)$$

(Amida ácida glicolica)

(glycolonitrile)

(glycolic acid amide)

The second process employs a proprietary formulation containing 41 % H2O2 with traces of catalyst and stabilisers called "Perox^gen Compounds" and 2 to 3 parts of 37 % formaldehyde solution to 1 part sodium cyanide and is stirred for one hour. The high temperature required for rapid reaction makes this process impractical for treating large volumes.

2.5 Technologies or advanced oxidation processes (TAO,PAO) to destroy cyanide

TAOs[18-22] are based on physicochemical processes capable of producing profound changes in the chemical structure of pollutants. The concept was initially established by Glaze and co-workers[19,23,24] , who defined OAPs as processes involving the generation and use of powerful transient species, primarily the hydroxyl radical (OH^{-}). This radical can be generated by photochemical means (including sunlight) or by other forms of energy, and is highly effective for the oxidation of organic matter. Some HATs, such as heterogeneous photocatalysis, radiolysis and other advanced techniques, also make use of chemical reductants that allow transformations on toxic pollutants that are not very susceptible to oxidation, such as metal ions or halogenated compounds .[25]

2.5.1. Heterogeneous photocatalysis

Heterogeneous photocatalysis[26] is a process based on the absorption of direct or indirect photons of light by a solid, which is a semiconductor, (heterogeneous photocatalyst, which is usually a broadband semiconductor) visible or UV, with sufficient energy, equal or higher than the gap energy of the semiconductor, Egap. Photocatalysis can be defined as the acceleration of a photoreaction by the presence of a catalyst. The catalyst, activated by the absorption of light, accelerates the process by interacting with the reactant through an excited state (C*) or by the appearance of electron-hole pairs, if the catalyst is a semiconductor (e^- - h +). In the latter case, the excited electrons are transferred to the reducible species, while the catalyst accepts electrons from the oxidisable species that will occupy the holes; in this way the net electron flow will be zero and the catalyst will remain unchanged .[27]

Within photocatalysis there are two types of techniques: Heterogeneous processes, which are mediated by a semiconductor as catalyst and homogeneous processes or processes mediated by ferric compounds, where the system is used in a single phase. The photon-absorbing species (C) is activated and accelerates the process by interacting with the other species in its new excited state (C*). In the case of heterogeneous processes, the interaction of a photon produces the appearance of an electron-hole (e- and h+), and the catalyst used will be a semiconductor. In this case, the excited electrons are transferred to the reducing species (Ox1) at the same time as the catalyst accepts electrons from the oxidising species (Red2), which occupies the hole spaces .[27]

2.5.1.1 Properties of titanium dioxide

TiO2 is an N-type semiconductor compound, it has 3 types of structures: anatase, rutile and brookite, anatase and rutile are the most used in photocatalytic reactions, the anatase phase shows higher activity[28] . The basic properties of anatase and rutile are shown in table 2.1.

Table 2.1. Properties of anatase and rutile TiO2 structures .[29]

	Rutile	Anatasa
Crystalline structure	Cubica	cubica
Specific gravity	4.2	3.9
Hardness (Mohs scale)	6.0-7.0	5.5-6.6
Dielectric coefficient (debye)	114	31
Refractive index	2.71	2.52
Energy gap (eV)	3.0	3.2
Melting point	1858	Transforms to rutile at 600°.

The structures of anatase and rutile TIO2 can be described in terms of octahedral TiO6 chains. The unit cell structures of rutile and anatase crystals are shown in Figures 2.3 and 2.4, each Ti ion^{+4} is surrounded by an octahedron of six O ions^{-2} . The octahedron in anatase is more distorted than in rutile, therefore, the anatase symmetry is lower than the orthorhombic one. The distance between two Ti atoms in anatase are larger (3.79 A and 3.04 A vs. 3.57 and 2.96 A in rutile) and the Ti-O distances are shorter in anatase than in rutile (1.934 and 1.980 A vs. 1.949 and 1.980 A in rutile). Each octahedron in the anatase structure is connected with eight neighbouring octahedra, in the rutile structure, there are ten neighbouring octahedra [29].

The above differences cause different specific gravities and electronic band structures between the two types of TIO2. The higher energy gap of anatase (3.2 eV) provides a better oxidation capability, therefore the anatase structure is more commonly used in photocatalysis.

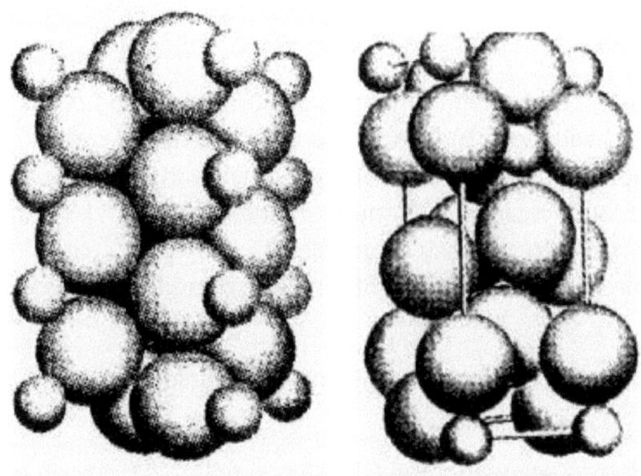

Figure 2.3 Rutile[29] . Figure 2.4 Anatase .[29]

2.5.2 Photocatalysis technology

The photocatalysis mechanism takes place when TIO2 nanoparticles are illuminated with ultraviolet (UV) rays, in which the UV photons excite the band containing the valence electrons and the electrons cross the conduction band, leaving gaps (h) in the valence band. For example, if the semiconductor (TIO2) is in an aqueous medium, the gaps will react with water molecules (H_2O) to produce hydroxyl radicals (OH^-), which are extremely oxidising and capable of oxidising cyanide to cyanate (CNO)$^-$[29] . Figure 2.5 illustrates the energy diagram and the photocatalytic process in a particle of TIO2.

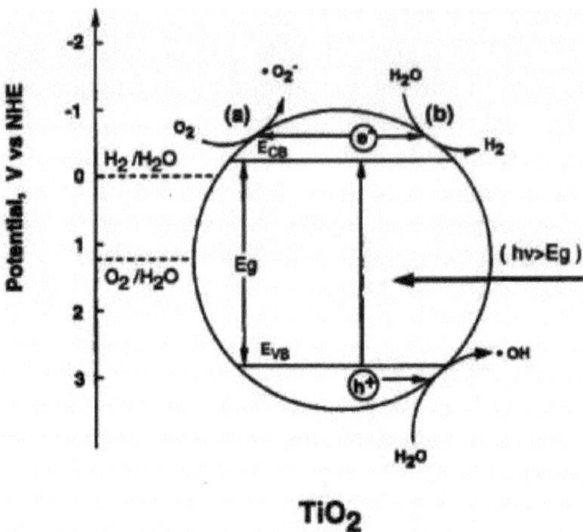

Figure 2.5 Schematic illustration of the energy correlation and redox mechanism on the TiO2 surface (a) with and (b) without the presence of oxygen .[30]

Augugliaro[30] reports that in an aqueous medium with cyanide, the first product of photocatalytic oxidation using TIO2 nanocrystals is CNO^- , the chemical reaction representing this oxidation process is as follows:

Formation of spaces with ultraviolet light (sunlight or artificial source);

$$TiO_2 + hv = h^+ + e^- \qquad (2.15)$$

Reaction of the valence band spaces (h) with the cyanide ion:

$$CN^- + 2\,h^+ + 2\,OH^- = CNO^- + H_2O \qquad (2.16)$$

Once this conversion is achieved, the CNO^- is completely oxidised and the final products are mainly CO_2y NO_3^- :

$$CNO^- + 4O_2 + 2OH^- + 3H_2O = CO^-_3 + NO^-_3 + 4H_2O_2 + e^- \qquad (2.17)$$

Finally, the reduction of oxygen consumes the electrons generated by the reaction (2.17) according to this chemical reaction:

$$O_2 + 2e^- + 2\,H_2O = H_2O_2 + 2\,OH^- \qquad (2.18)$$

In another study for the photocatalytic degradation of ferrocyanide[31] , the absorption band of TiO2 was found to be 350 nm and the degradation of this complex was carried out within 10 hours at a pH range of 810 with destruction efficiencies above 85%.

This technology has not been successfully applied industrially due to the cost of TiO2. This problem is solved by recovering and recycling TiO2 microelectrodes. However, current filtration techniques have proven to be very poor at recovering the nanometric TIO2 particles and then recycling them to the photocatalytic cyanide destruction process.

To successfully overcome this deficiency, the electrocoagulation process was used to recover the TIO2 nanoparticles and subsequently recycle them to the cyanide destruction process. To separate the TIO2 particles from the iron hydroxides, leaching with sulphuric acid is used by the following reaction:

$$[Fe(OH)_2.TiO_2] + H_2SO_4 = FeSO_4 + 2H_2O + TiO_2 \quad (2.19)$$

2.5.3 Recent photocatalysis research

Photocatalytic oxidation of cyanide has been proposed as a viable option for cyanide treatment[32] and an alternative to conventional chemical methods. Other authors reported photodegradation of cyanide in industrial wastewater using TiO2 as a catalyst[33] and quantitatively examined the CO_2 evolution, O_2 consumption, N2 evolution, and the

formation of the intermediate ion CNO^-. Also the photocatalytic oxidation of cyanide using Degussa P-25 as catalyst was investigated by other authors[33] where they confirmed that cyanide is first oxidised to cyanate and subsequently to nitrite and nitrate.

In this study, the catalytic activity of Degussa P-25 and TiO2 (rutile) was examined. The irradiation was carried out at laboratory scale, the 500 ppm CN solution⁻ was photodegraded using TiO2 in an alkaline medium with an initial pH of 12. The effect of different amounts of TiO2 for 500 ppm solution was then also reviewed. Table 2.2 shows a comparison of different articles where the oxidation of cyanide using the photocatalysis method is used.

Table 2.2 Comparison of results obtained for cyanide oxidation using heterogeneous photocatalysis

heterogeneous photocatalysis.

Article Reference	Conditions of Operation	% Reduction CN^-
[34]	0.15 g/L TiO_2 400 ppm CN initial 450 W xenon lamps 0.1 M KCN solutions Time 30 min. with xenon lamps and 2 days with sunlight Doped and undoped Anatase	84 % with xenon lamp 30 min. exhibition 99% with sunlight and 2 days of exposure
[35]	0.25 g/L of TiO2, 200 ppm initial CN radiation power 150 W hydrogen peroxide (30% in water m/m) pH 9.5,	85 % With recirculation time of 2 hrs. 97% Recirculation 4 hrs. and re-injecting H_2O_2
[36]	0.40 g/L TiO_2 20 g CN per kWh of solar energy supplied CPCS-based solar technology	Complete degradation
[37]	25 mg/L TiO_2 100 ppm initial CN 120 min. exposure with UV light pH= 10.5	95 % after 8hrs sun exposure 99 % after 2 hr. UV light
[38]	0.50 g/L TiO_2 100 mg/dm^3 initial CN 1 hr exposure time with UV pH = 12	97 % for suspended anatase 75 % anatase supported

As can be seen in this table, the main limitation for the use of this oxidation method has been the recovery of TiO2 nanoparticles, which due to their extreme fineness cannot be efficiently recovered by normal filtration methods, Therefore, to remedy this weakness in this technology, electrocoagulation technology was used in this study to more efficiently recover the TiO2 particles and recycle them to photodegradation technology for cyanide destruction.

2.6 arsenic removal from industrial effluents

2.6.1 Arsenic

Arsenic is found in many allotropic forms and has both metallic and non-metallic properties. Arsenic occurs naturally in sedimentary and volcanic rocks (it forms 0.00005 % of the earth's crust) and also in geothermal waters[39] . In nature it occurs most frequently in the form of arsenic sulphide (As_2S_3) and arsenopyrite (FeAsS), usually found as impurities in mining deposits. The chemical properties of arsenic are shown in Table 2.3.

Table 2.3 Chemical properties of arsenic .[39]

Symbol	As
Classification	Nitrogenoid element, group 15, metalloid
Atomic number	33
Oxidation numbers	-3,0,+3,+5
Isotopes	1 natural isotope (As)^{75} 32 unstable (half-life ranges between 0.09577 sec. As^{66} and 80.3 days, As^{73}

2.6.2 Arsenic in water

In water (surface and groundwater) arsenic is commonly found in the +5 (arsenate) and +3 (arsenite) oxidation state. In surface water with high oxygen content, the most common species is pentavalent arsenic or arsenate (As^{+5}). Under reducing conditions, usually in lake sediments or groundwater, trivalent arsenic or arsenite (As^{+3}) predominates .[39]

Arsenite is found in solution as H_3AsO_3, $H_2AsO_3^-$, AsO4 and $H_2AsO_4^2$, in natural waters with pH between 5 and 9. Arsenate is found in stable form in waters with high oxygen levels as H_3AsO_4 in a pH range of 2 to 13. The conversion of As^{+3} to As^{+5} or As^{+5} to As^{+3} is quite slow. Reduced As^{+3} compounds can be found in oxidised media and oxidised As^{+5} compounds in reduced media .[40]

2.6.3 Toxicity

The toxicity of arsenic depends on its oxidation state, chemical structure and solubility in biological media. The scale of arsenic toxicity decreases in the following order:

Arsine or hydrogen hydride (AsH_3) > inorganic As^{+3} > organic As^{+3} > inorganic As^{+5} > organic As^{+5} > arsenical compounds and elemental arsenic .[40]

The toxicity of As^{+3} is 10 times higher than that of As^{+5} and the lethal dose for adults is 1- 4 mg As/kg per person. The United States Environmental Protection Agency, USEPA, classifies arsenic as a Group A carcinogen because of evidence of adverse health effects. Exposure to 0.05 ppm may cause 31.33 cases of skin cancer per 1 000 population and has considered lowering the maximum acceptable limit from 0.050 ppm to 0.010 - 0.020 ppm. The International Agency for Research on Cancer has classified it in Group I because they have sufficient evidence of carcinogenicity to humans .[41]

Natural elimination from the human body is via urine, faeces, sweat and skin epithelium (desquamation)[41] . Some studies of arsenic toxicity indicate that many of the current standards based on WHO guidelines are too high, and suggest the need to re-evaluate the limit values based on epidemiological studies; for example, in Taiwan they estimate that the limit value should be reduced from 0.02 to 0.0005 ppm, in other cases it seems that these values should be increased, according to regional conditions. In Latin America it has been observed that at similar levels of arsenic in different conditions (climatic, nutritional and other), the level of effects is different .[42]

2.6.4 Arsenic Removal Methods from Industrial Effluents

In the last 50 years, the contamination of the environment has been seriously affected by the emission of heavy metals (As, Pb, Cr, Hg, etc.). In the case of chemical effluents, the control of these pollutants has been carried out using conventional pressurised air technology (bubble columns, packed beds and flotation cells), precipitation processes, cementation, solvent extraction processes and processes using selective resins whose technology requires high investment costs and large installation space which does not make it very attractive, as well as dependence and payment of patent fees for foreign technology. Table 2.4 shows different technologies currently used for the treatment of arsenic-contaminated water. Of these processes, the most economical and recommended method for the removal of heavy metals such as arsenic from chemical effluents is chemical precipitation; a process whereby the metal species of a soluble phase (usually the ionic residue of arsenic) are removed from solution by reacting with a precipitating agent which is added to produce an insoluble component.

Table 2.4. Technologies used for arsenic removal from water .[45]

Technology	Advantages	Disadvantages	Elimination
Oxidation-Precipitation.			
With air.	Simple and low-cost technology.	Slow process.	80%
Chemistry.	Relatively simple and quick process.	Specific pH control in each tank.	90%
Coagulation - Precipitation.			
Coagulation with Alumina.	Simple and low-cost operation.	Produces toxic sludge.	90%
Coagulation with Iron .	More efficient than alumina.	Requires large tanks.	95%
Precipitation with lime.	It uses common chemicals.	Requires pH adjustment in each tank.	91%
Adsorption technologies.			
Activated alumina.	Commercially available technology.	Replace after 4-5 regenerations.	88%
Iron mixed with sand.	Cheap reagent.	Produces toxic waste.	93%
Lonium Exchange Resins.	Process less dependent on the pH of the water.	It requires high-tech operation and maintenance.	92%
Membrane Processes.			
Nanofiltration.	High removal efficiency.	High cost of capital.	95%
Reverse Osmosis.	High removal efficiency.	Requires specialised personnel.	96%
Electrodialysis.	High removal efficiency.	Produces toxic	95%

		wastewater.	

In this originating solid component, the contaminating metal is captured. The metals precipitate at various pH levels, depending on the metal ion, and form an insoluble salt. Subsequently, the precipitate can be separated from the wastewater using a physical separation process such as sedimentation or filtration; however, arsenic removal efficiencies with these removal processes are less than 85% .[43]

Due to the above, the electrochemical process of Electrocoagulation was used, which does not require the addition of chemicals for arsenic removal and most importantly is a fast, compact process that does not generate polluting by-products .[44]

2.7 Electrocoagulation technology

The principle of electrocoagulation is based on the coagulation between the polyvalent cations formed by the electroKtic oxidation of the iron and aluminium anodes, and the contaminants or compounds in the aqueous medium, such as TiO2. In this process, the electrophoretic movement tends to concentrate the negatively charged particles in the anode region and the positively charged ions in the cathode region.

Consequently, the ions released by the electrodes neutralise the charged particles with opposite charges, thus facilitating the coagulation of the resulting species; due to the simultaneous generation of gases during electrolysis (production of H_2 and O_2), the agglomerate or coagulum containing the TiO_2 floats, eliminating or recovering these species from the effluent[8] . The difference between chemical coagulation and electrocoagulation is the origin of the coagulant, since in electrocoagulation the cation comes from the dissolution of the metal anode, either iron or aluminium.

Electrocoagulation is a process frequently used during the last decade in the United States and Europe for the treatment of industrial effluents with toxic substances[44] . This technology is used in the textile industry for the removal of colours from effluents[46] , treatment of oils[47] , removal of grease from restaurants[48] , removal of suspended particles from effluents[49] and also for the treatment of toxic metals from industrial effluents[50] . In most of these studies, the parameters used have been determined empirically and lack a chemical and physical basis for the process. For this reason, it is necessary to carry out a basic study to understand the fundamental mechanisms of the electrocoagulation process.

2.7.1 Reactions in the electrocoagulation process

The mechanism of electrocoagulation is highly dependent on the chemistry of the aqueous medium, especially the conductivity. Other factors such as pH, particle size and concentration of the chemical reagents also influence the process. Ion removal by EC can be explained by the following general reactions .[51]

At the aluminium electrode, aluminium is electroKtically dissolved to produce monomeric cationic species, such as Al^{3+} and $Al(OH)_2$, according to the following reactions:

Al ആ $Al^{3+}(aq) + 3e^-$ (2.20)

$Al^{3+}(aq) + 3H_2O$ ൌ $Al(OH)_3 + 3H^+(aq)$ (2.21)

$n\ Al(OH)_3$ ൌ $Al(OH)_3$ (2.22)

However, depending on the pH of the aqueous medium, other species may be present in the system, such as $Al(OH)^{2+}$, $Al_2(OH)\ _2^{4+}$ y $Al(OH)\ _4^-$. Under appropriate conditions, multimer species of aluminium hydroxides can form whose cationic gel complexes can remove contaminants by adsorption, to produce a neutralisation of their charges and form a precipitate .[51]

At the iron anode, the oxidation of iron in an electroKtic system produces iron hydroxide, $Fe(OH)_n$, where n =2 or 3. Two mechanisms have been proposed for the production of this compound by the EC process:

<u>Mechanism 1</u>

Anode:

$4Fe_{(s)}$ ണ്ട $4Fe^{+2}{}_{+(aq)} + 8e^-$ (2.23)

$4Fe^{2+}{}_{(aq)} + 10H_2O_{(l)} + O_{2(g)}$ ണ്ട $4Fe(OH)_{3(s)} + 8H^+{}_{(aq)}$ (2.24)

Cathode: $8H^+{}_{(aq)} + 8e^-$ ണ്ട $4H_{2(g)}$ (2.25)

Global: $4Fe_{(s)} + 10H_2O_{(l)} + O_{2(g)}$ ണ്ട $4Fe(OH)_{3(s)} + 4H_{2(g)}$ (2.26)

<u>Mechanism 2</u>

$Fe_{(s)}$ ണ്ട $Fe^2{}_{+(aq)} + 2e^-$ (2.27)

Anode: $Fe^{2+}{}_{(aq)} + 2OH^-{}_{(aq)}$ ണ്ട $Fe(OH)_{2(s)}$ (2.28)

Cathode: $2H_2O_{(l)} + 2e^-$ ണ്ട $H_2(g) + 2OH^-{}_{(aq)}$ (2.29)

Global: $Fe_{(s)} + 2H_2O_{(l)}$ ണ്ട $Fe(OH)_{2(s)} + H_{2(g)}$ (2.30)

The $Fe(OH)_n$ formed remains in the aqueous solution as a gel-like suspension, which can remove the pollutants from the effluent by complexation or electrostatic attraction, followed by coagulation. One hypothesis is that the mode of complexation at the contaminant surface occurs when the contaminant acts as a ligand to form a chemical bond with the iron hydroxides. The pre-hydrolysis of Fe cations^{3+} by EC allows the formation of reactive bonds for wastewater treatment .[51]

2.7.1.1 Electrocoagulation reactions at different pHs

The mechanism of electrocoagulation with iron electrodes at different pH was proposed by Hector A. Moreno .[52]

For pH <4

Anode:

$Fe \rightarrow Fe^{+2} + 2e^-$ (2.31)

$Fe \rightarrow Fe^{+3} + 3e^-$ (2.32)

Catodo:

$2H^+ + 2e^- \rightarrow H_{2(g)}\uparrow$ (2.33)

Electrochemicals depend on kinetics and thermodynamics. EC can be considered as an accelerated corrosion process. The rate of the reaction will depend on the elimination $[H^+]$ via evolution of H_2 .52]

For pH 4<pH<7

Anode: Reactions (2.31) and (2.32)
Comments: Iron also undergoes hydrolysis.

$$Fe + 6H_2O \rightarrow Fe(H_2O)_4(OH)_{2\,(aq)} + 2H^{+1} + 2e^{-1} \quad (2.34)$$

$$Fe + 6H_2O \rightarrow Fe(H_2O)_3(OH)_{3\,(aq)} + 3H^{+1} + 3e^{-1} \quad (2.35)$$

Fe(III) hydroxide begins to precipitate with a yellow floc.

$$Fe(H_2O)_3(OH)_{3\,(aq)} \rightarrow Fe(H_2O)_3(OH)_{3\,(s)} \quad (2.36)$$

$$2Fe(H_2O)_3(OH)_3 \leftrightarrow Fe_2O_3\,(H_2O)_6 \quad (2.37)$$

Catodo:

$$2H^{+} + 2\,e^{-} \rightarrow H_2(g)\uparrow \quad (2.38)$$

For pH 6<pH<9
Anode: Reactions (2.31) and (2.32).
Precipitation of Fe(III) hydroxide continues, and precipitation of Fe(II) hydroxide also occurs, presenting a dark green floc.
The pH for the minimum solubility of $Fe(OH)_n$ is in the range of 7-8. The electrocoagulated floc is formed due to the polymerisation of iron oxyhydroxides.

$$Fe(H_2O)_4(OH)_{2\,(aq)} \rightarrow Fe(H_2O)_4(OH)_{2\,(s)} \quad (2.39)$$

$$2Fe(OH)_3 \leftrightarrow Fe_2O_3 + 3H_2O \text{ (hematita, magemita)} \quad (2.40)$$

$$Fe(OH)_2 \leftrightarrow FeO + H_2O \quad (2.41)$$

$$2Fe(OH)_3 + Fe(OH)_2 \leftrightarrow Fe_3O_4 + 4H_2O \text{ (magnetita)} \quad (2.42)$$

$$Fe(OH)_3 \leftrightarrow FeO(OH) + H_2O \text{ (goetita, lepidocrocita)} \quad (2.43)$$

Hematite, magemite, magnetite, lepidocrocite and goethite have been identified by EC by Parga[53] , and Gomes .[53]

Cathode $$2H^{+} + 2\,e^{-} \rightarrow H_{2(g)}\uparrow \quad (2.44)$$

The evolution of hydrogen takes place in a major way but the $[H^{+}]$ now comes from the hydrolysis of iron and weak acids .[52]
Other reactions include:

$$Fe + 6H_2O \rightarrow Fe(H_2O)_4(OH)_{2\,(s)} + H_{2(g)}\uparrow \quad (2.45)$$

$$Fe + 6H_2O \rightarrow Fe(H_2O)_3(OH)_{3\,(s)} + 1\tfrac{1}{2}\,H_{2(g)}\uparrow \quad (2.46)$$

The conditions in the whole cell are not constant concentrations, species and pH are changing. This can be illustrated with a Pourbaix diagram of Fe. The process appears to occur in the region parallel to the hydrogen evolution kinetics and conditions change to the right as shown in Figure 2.6 .[52]

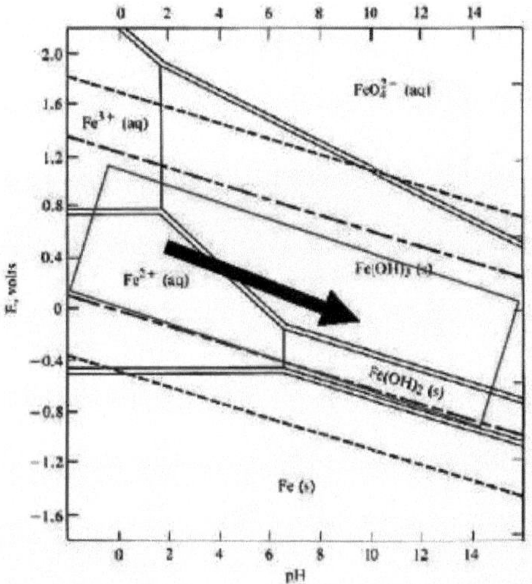

Figure 2.6. Pourbaix diagram for Fe, showing the region and direction in which electrocoagulation proceeds .[53]

Accordingly, the electrocoagulation process is carried out on the principle that coagulation is carried out with the ions produced electrokinetically from the iron and aluminium anodes which adsorb the pollutants from the aqueous medium containing the species to be removed or recovered. The general idea is to use the production of polyvalent cations formed by the oxidation of the anodes (Fe and Al) and the gases produced during electrolysis (H_2 and O_2) for the removal of nanometric particles (TIO2) and dissolved heavy metals (arsenic).

In this process, the electrophoretic movement tends to concentrate the negatively charged particles in the anode region and the positively charged ions in the cathode region. Consequently, the ions released by the electrodes neutralise the oppositely charged particles, thereby facilitating coagulation. Figure 2.7 illustrates the schematic diagram of the above process.

During the process, the bubbles produced by the gases formed by electrolysis can adsorb the particles or coagulates containing TIO2 and arsenic, which are transported to the top of the reactor where they are concentrated and removed. Also metal ions can react with the produced OH ions$^-$ at the cathode where hydrogen evolution occurs and forms an insoluble hydroxide which adsorbs the pollutants from the solution and also contributes to coagulation due to neutralization of the colloidal particles of TIO2 and arsenic. This particle agglomeration behaviour is also due to the influence of van der Walls forces of attraction .[53]

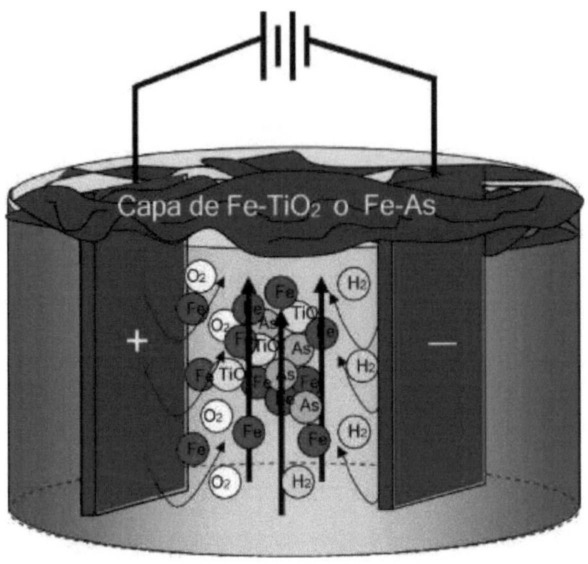

Figure 2.7. Schematic diagram of the electrocoagulation process .[53]

In general, this process has the following advantages: The treated effluents are free of impurities such as colour, odour, and solids; since an electric field is applied, colloidal particles can be removed, thus facilitating the clarification of the aqueous phase; the treatment unit is not very bulky and maintenance is very easy since it has no moving parts; finally, this technology can be used in remote areas of the country where there are electricity problems, since solar energy can be used to carry out this process.

The following is a comparison of the CE process against biological and chemical treatment:

- Electrocoagulation v/s biological treatment :[54]

- The electrocoagulation system applied to wastewater, compared to conventional biological systems, requires less surface area (50 to 60% less).
- Electrocoagulation residence times are 10 to 60 seconds, compared to biological systems that require 12 to 24 hours.
- They are compact units, easy to operate, with lower energy consumption and (more compact) sludge production than conventional biological systems.
- The electrocoagulation cells are built in FVR and are installed on the ground. Therefore, they do not require major civil works, like chemical and biological systems.
- Investment costs are 50% lower than for biological systems.
- Electricity consumption per m3 of treated water (between 0.1 to 1 Kwh/m^3) is lower than in conventional treatment systems.
- They are 100% automatic units, which are used when required, with response times of 10 to 60 seconds, at their level of efficiency.

- Electrocoagulation v/s chemical treatment :[54]

- They are compact units, easy to operate, with lower energy consumption and (more compact) sludge production than conventional chemical treatment systems.

- The electrocoagulation cells are built in FVR and installed on the ground. Therefore, they do not require major civil works, like chemical and biological systems.
- Electricity consumption per m3 of treated water, between 0.1 to 1 Kwh/m^3 , is lower than in conventional treatment systems.
- Operation costs are 50% to 60% lower than chemical systems.
- Adaptable to different types of processes.

The electrocoagulation process is a novel process in which the kinetic and thermodynamic mechanism of adsorption of TЮ2 nanoparticles and arsenic ions on magnetite and goethite species generated on the iron electrodes used in the EC reactor will be studied.

2.7.2 Current density

The current density is the ratio of the current supplied to the electrodes to the area of the electrodes, calculated with the formula :[55]

$$D = \frac{I}{A} \qquad (2.47)$$

Where:

D = Current density (Amp/cm $)^2$

I= Applied current (amperes)

A = Area of the electrodes (cm $)^2$

2.7.3 Electrode Disolution

The amount of metal that is dissolved or deposited depends on the amount of electricity passing through the electrocoagulation solution and the residence time of the water in the electrocoagulation cell.

A simple relationship between the current density (A cm.$^{-2}$) and the amount of substance (M) dissolved (g of M per cm.$^{-2}$) is derived from Faraday's law :[55]

$$W = \frac{(D * t * M)}{nF} \qquad (2.48)$$

Where:

W = Amount of dissolved electrode material (grams per cm. $)^{-2}$

D = Current density (A per cm $)^{-2}$

t = Time (in seconds).

M = Relative molar mass of the electrode.

n = Number of electrons in the redox reaction.

F = Faraday constant (96,500 columbia)

2.7.4 Energy consumption

The energy consumption of an electrocoagulation reactor can be calculated with the equation :[55]

$$E = \frac{(I * v * t)}{1000} \qquad (2.49)$$

Where:

E= Energy (kW^hr)

I = Current (Amperes) v = Load (volts) t = Time (hrs.)

2.7.5 Cost of treatment

The cost of the treatment of the energy supplied is calculated with the following formula :[55]

$C = E^{*}Ce$ (2.50)

Where:
C = Cost of treatment
E = energy (Kw.hr)
Ce = Cost of 1 kW.hr of energy
2.8 Thermodynamic analysis
2.8.1 Langmuir Isotherm
In 1918 I. Langmuir deduced the Type I isotherm (corresponds to a monolayer adsorption, it is the characteristic isotherm of a physisorption or chemisorption process[56]) using a simplified model of the surface of a solid:

- The surface provides a certain number of positions for adsorption and all are equivalent.
- Only one molecule is adsorbed on each position.
- Their adsorption is independent of the occupation of neighbouring positions (the adsorbed molecules do not interact with each other sQ.

The dynamic adsorption process can be thought of as :[56]

$$A_{(liq)} + M_{(sol)} \underset{Kd}{\overset{Ka}{\longleftrightarrow}} A - M_{(sol)} \quad (2.51)$$

ka: rate constant for adsorption kd: rate constant for desorption.
In order to express the extent of adsorption, the coating fraction 0 is entered. Taking into account that only one molecule can be adsorbed on each position :[56]

$$\Theta = \frac{\text{no. moléculas adsorbidas}}{\text{no. posiciones adsorción}} = \frac{\text{no. posiciones ocupadas}}{\text{no. posiciones adsorción(N)}} \quad (2.52)$$

no. adsorbed motecules no. adsorption positions
no. positions occupied no. positions adsorption(N)

Where N is the total number of adsorption positions on the surface.
At an instant t is fulfilled:
-number of occupied adsorption positions = cN
-number of free adsorption positions $= N - \Theta N = N(1-\Theta)$
If we consider first order kinetics with respect to each member, we obtain that the adsorption rate is proportional to the number of collisions between the molecules of the liquid phase and the empty adsorption positions, since only a monolayer is formed:

$$V_a = k_a C(1-\Theta)N \quad (2.53)$$

The desorption rate will be proportional to the number of adsorbed molecules:

$$V_d = K_d N\Theta \quad (2.54)$$

The two velocities are equalised at equilibrium, from which we obtain:

$$k_a CN(1-\Theta) = k_d = N\Theta \quad (2.55)$$

$$k_a P - k_a C\Theta = k_d \Theta \quad (2.56)$$

If we clear the covering fraction:

$$\Theta = \frac{k_a C}{k_d + k_a C} \qquad (2.57)$$

Defining the equilibrium constant as $K = ka/kd$ gives the Langmuir isotherm:

$$\Theta = \frac{KC}{1 + KC} \qquad (2.58)$$

Alternatively, this expression can be deduced from the equilibrium between products (adsorbed molecules) and reactants (free positions and gas molecules):

$$K = \frac{N(1-\Theta)}{N\Theta C} \qquad (2.59)$$

The constant K = ka /kd determines the equilibrium state at a given pressure. This relation gives rise to the well-known Langmuir isotherm, which tends to fit the experimental data better than the Freundlich isotherm .[56]

2.8.1.1 Separation factor

A characteristic of the Langmuir isotherm can be expressed in terms of a dimensionless constant called the separation factor or equilibrium parameter R_L, which is used to predict whether an adsorption system is favourable or unfavourable. The separation factor is defined by the following equation :[56]

$$R_L = \frac{1}{1 + bC_0} \qquad (2.60)$$

0

Where C_o is the initial concentration of the adsorption system and b is the Lagmuir constant, the isotherm is unfavourable if $R_L > 1$, the isotherm is linear if $R_L = 0$, the isotherm is favourable when $0 < R_L < 1$ and the isotherm is irreversible when $R_L = 0$.

2.8.1.2 Other Adsorption Isotherms

As mentioned above, the Langmuir isotherm tends to fit the experimental data better than other isotherms, such as Freundlich and DR (Dubinin-Radushkevich). These isotherms are presented below for a comparison of data fits. The isotherms are presented in normal and linear form.

2.8.1.1.2.1 Freundlich Isotherm

The Freundlich equation is an empirical equation used to describe heterogeneous systems, which is characterised by the heterogeneity factor 1/n[57] , therefore the empirical equation can be written:

$$N = K_F C_e^{1/n} \qquad (2.61)$$

Where N is the concentration of the solid phase (mg/g), C_e is the concentration of the liquid phase (mg/L), K_F is the Freundlich constant and 1/n is the heterogeneity factor. The linear form of the Freundlich expression is obtained by taking logarithms of equation (2.61):

$$\ln N = \ln K_F + \frac{1}{n} \ln C_e \qquad (2.62)$$

Therefore, by plotting ln N vs C_e one can determine the KF constant and the exponent 1/n, which are Freundlich constants and represent the adsorption capacity and the adsorption intensity respectively. When n>1 represents favourable adsorption conditions .[57]

2.8.1.2.2 D-R Isotherm (Dubinin-Radushkevich)

The D-R isotherm is more general than the Langmuir isotherm, because it does not assume a homogeneous surface or constant sorption potential. The equation is:

$$N = N_{max} \exp(-K\varepsilon^2) \qquad (2.63)$$

The linear form of the D-R isotherm is obtained by taking logarithms of equation (n) and is shown in the following equation:

$$\text{Ln } N = \ln N_{max} - K\varepsilon^2 \qquad (2.64)$$

Where N is the amount adsorbed, K is a constant related to the adsorption energy, N_{max} is the theoretical saturation capacity, e is the Polanyi potential equal to RT ln $(1+1/Ce)$.[57]

2.8.2 Number of moles adsorbed per gram of adsorbent

From the difference in initial and final concentrations of TIO2 or arsenic, the volume of solution used and the mass of dissolved iron hydroxides (m_c), N can be calculated, which is the number of moles adsorbed per gram of adsorbent .[58]

The number of moles adsorbed per gram of adsorbent is calculated from the equation:

$$N = V * \frac{Co - C}{m_c} \qquad (2.65)$$

Where:

N= Number of moles adsorbed per gram of adsorbent

V = Volume of the solution

Co = Initial concentration

C = Final concentration

m_c= Mass dissolved hydroxide

2.8.3 Calculation of Nmax and Langmuir constant (K)

If we assume that Nmax is the maximum amount of adsorbate that can be adsorbed on one gram of iron hydroxide, the degree of coverage 0 is 0 = N/Nmax. Under these conditions, the Langmuir isotherm (equation 2.58) can be rewritten as follows:

$$N = \frac{N_{max} KC}{1 + KC} \qquad (2.66)$$

And rearrange as:

$$\frac{C}{N} = \frac{C}{N_{max}} = \frac{1}{KN_{max}} \qquad (2.67)$$

If the system follows the behaviour described by the Langmuir isotherm, the graph of the C/N ratio as a function of the equilibrium concentration C should give a straight line of slope 1/Nmax and ordinate to the origin 1/KNmax .[58]

2.8.4 Specific BET area (Brunauer, Emmet and Teller)

The BET theory is a well-known technique for the physical adsorption of a gaseous molecule on a solid surface, which is the basis for an important analytical technique for the measurement of the specific surface area of a material.

The concept of the theory is an extension of Langmuir's theory, which is the basis of monolayer molecular adsorption to multilayer adsorption with the following assumptions: a) Gas molecules adsorb physically on an infinitely layered solid, b)

There is no interaction between each adsorption layer and c) Langmuir's theory can be applied to each layer. The resulting BET equation is the following :[59]

$$\frac{1}{V[(Po/P)-1]} = \frac{C-1}{VmC}\left(\frac{P}{Po}\right) + \frac{1}{VmC} \quad (2.68)$$

P and Po are at equilibrium and the saturation pressure of the adsorbate at the adsorption temperature, V is the amount of gas adsorbed (e.g. in volume units e) and Vm is the amount of gas adsorbed on the monolayer C is the BET constant, expressed by:

$$C = \exp\left(\frac{E_1 - E_L}{RT}\right) \quad (2.69)$$

E1 is the heat of adsorption for the first monolayer and EL is for the second and upper layers and is equal to the heat of liquefaction.

2.8.4.1 BET Chart

Equation (2.68) is an adsorption isotherm and can be plotted as a straight line with $1/v[(P_0/P)-1]$, on the y-axis $\varphi = P/P_0$, on the x-axis according to the experimental results. This graph is called the BET graph (Figure 2.8). The linear relationship of this equation holds only in the range of 0.05 $<P/P_0<$ 0.35.

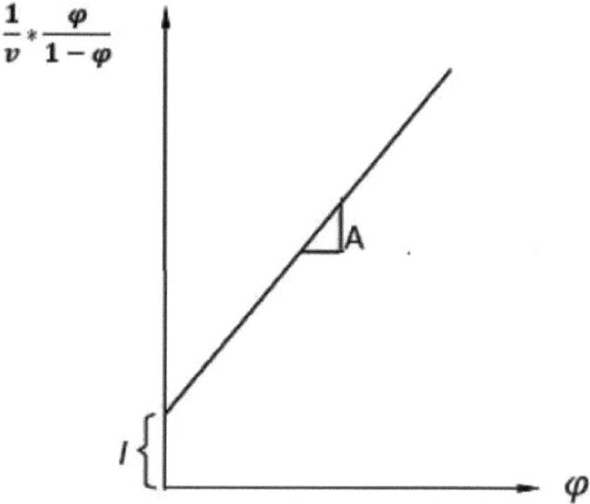

Figure 2.8. BET chart .[59]

The value of the slope A and the ordinate to the origin y of the line is used to calculate the amount of gas adsorbed on the monolayer Vm and the BET constant c. The following equations are used.

$$Vm = \frac{1}{A+I} \quad (2.70)$$

$$C = 1 + \frac{A}{I} \quad (2.71)$$

The BET method is widely used for the calculation of surface areas of solids by physical adsorption of gas molecules. The total surface area S_{total} and the specific surface area S are evaluated with the following equations :[59]

$$S_{BET,total} = \frac{(VmNs)}{V} \qquad (2.72)$$

$$S_{BET} = \frac{S_{total}}{a} \qquad (2.73)$$

N : Avogadro number
S: adsorption cross section
V: Molar volume of the adsorbent gas
A: Molar weight of the adsorbed species
The usual value for adsorbents consisting of small, porous particles is between 10 and 1,000 m /g.[2]
2.8.5 Thermodynamics of the reaction
Using the following relationships the thermodynamic parameters of the adsorption process were calculated:

$$\Delta G^{\circ} = -RT\ln b \qquad (2.74)$$

$$\Delta H^{\circ} = -RT\ln b - b_0 \qquad (2.75)$$

$$\Delta S^{\circ} = \frac{\Delta H^{\circ} - \Delta G^{\circ}}{T} \qquad (2.76)$$

Where AH° is the enta^a change, AS° is the entrop^a change, AG is the gibbs free energy change, b and b_0 are Langmuir constants at final and initial concentration. The negative value of the free energy suggests the feasibility of the process in both cases. The negative value of the enthalpy confirms the exothermic nature of the process, furthermore the positive value of the entropy change shows the increasing randomness of the process .[60]
2.9 Kinetic analysis
2.9.1 Langmuir-Hinshelwood model
The Langmuir-Hinshelwood kinetic model predicts the reaction rate behaviour with the following equation :[61]

$$r = -\frac{dc}{dt} = \frac{kKC}{KC+1} \qquad (2.77)$$

Where k represents the kinetic constant of the reaction, K the cyanide adsorption equilibrium constant and C the concentration of TiO2 or AS[61] . This equation can be linearised to fit the experimental data to the model and find their respective constants (table 2.5 and figure 2.9).

$$-\frac{dt}{dc} = \frac{1}{k} * \frac{1}{C} + \frac{1}{kK} \qquad (2.78)$$

The constants k and K are calculated from:

$$y = mx + b \quad (2.79)$$

$$m = 1/kK \quad (2.80)$$

$$b = 1/k \quad (2.81)$$

$$k = 1/b \quad (2.82)$$

$$K = b/m \quad (2.83)$$

When the kinetic constant (k) is greater than the adsorption constant (K), it means that the controlling phenomenon for the recovery or removal process of the compounds is the rate of adsorption of the pollutants (adsorption stage 2) on the catalyst surface rather than the rate of conversion of the pollutants (stage 3 of the adsorption process)[61] .

Table 2.5. Example of experimental data to plot in the Langmuir- Hinshelwood model [61].

Time (min)	$[CN^-]$ ppm	1/C	dt/dC
0	405.74	0.00246463	0.28143884
15	347.56	0.0028772	0.32764652
30	309.46	0.00323144	0.36732077
45	275.54	0.00362924	0.41187456
60	245.34	0.00407598	0.46190933
90	180.02	0.00555494	0.62755309
120	139.98	0.00714388	0.8055143

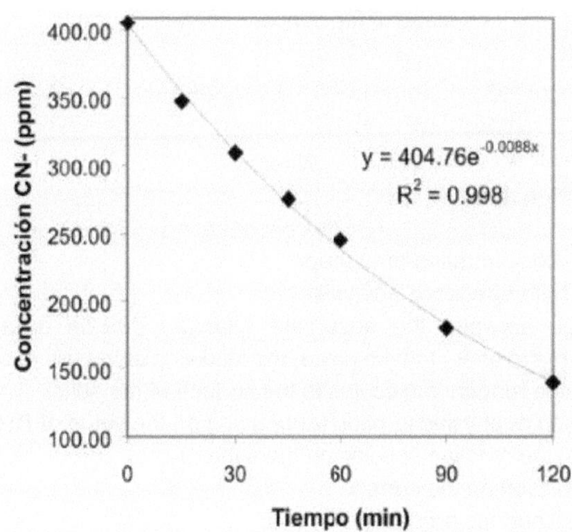

Time (min)

Figure 2.9 Langmuir -Hinshelwood model graph of Concentration CN vs. time .[61]

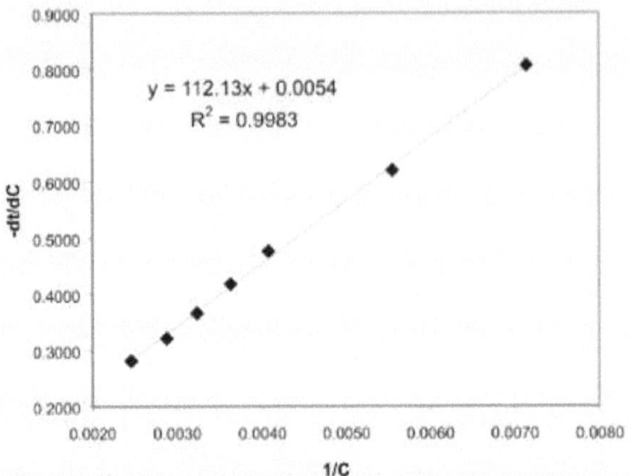

Figure 2.10 Langmuir-Hinshelwood model, graph of -dt/dC vs 1/C .[61]

The reaction velocities calculated from the first-order kinetic model are:

TiO_2 reduction rate :[61]

$$-r = \frac{-dC_{TiO2}}{dt} = m * C_{TiO2} \quad (2.84)$$

Where:

- r = reduction rate of TiO_2

C_{TiO2} = concentration of TiO_2

m = slope of the concentration-time graph

As reduction rate:

$$-r = \frac{-dC_{As}}{dt} = m * C_{As} \quad (2.85)$$

Where:

- r = As reduction rate

C_{As} = As concentration

m = slope of the concentration-time graph

2.9.2 Stages of heterogeneous analysis

For the reaction between the adsorbate (titanium dioxide or arsenic) and the adsorbent (iron hydroxides) to take place, the following steps are necessary :[62]

1) Diffusion of the reagent molecules to the surface of the solid.
2) Chemisorption or physiadsorption (depending on the value of ДH of the system) of at least one of the reactive species on the surface
3) Chemical reaction on the surface
4) Desorption of surface products
5) Diffusion of the products into the fluid phase

The slowest of these processes will determine the rate of reaction. Diffusion steps 1 and 5 depend on temperature, pressure, and adsorbate concentration and are generally fast, so any of steps 2, 3 or 4 can be the limiting (slowest) step, Langmuir assumed that step 2 or 3 of the surface reaction is the slowest step in the process, to know which of these steps is the limiting step it is necessary to know the values of the kinetic and adsorption constants, when the kinetic constant is greater than the adsorption constant, the controlling phenomenon or slowest stage of the process is the adsorption (chemisorption or physiadsorption) on the surface of the adsorbent (step 2). When the adsorption constant is higher than the kinetic constant, the slowest stage is the chemical reaction on the surface (stage 3) .[62]

III EXPERIMENTATION

In this chapter, experiments were carried out to perform a thermodynamic and kinetic study of the adsorption of titanium dioxide and arsenic on iron species generated by electrocoagulation. The first step was the oxidation of cyanide using the photocatalysis process with titanium dioxide nanoparticles, Secondly, electrocoagulation tests were carried out to recover the titanium dioxide from the photocatalytic oxidation and lastly, arsenic removal tests were carried out using the electrocoagulation process. The electrocoagulation tests were then carried out in two parts, the first part to carry out the adsorption and thermodynamic study and the second part to carry out the kinetic study.

III.1 Cyanide determination

3.1.1 Potentiometric analysis

This method is suitable for concentrations below 100 ppm and is based on the creation of an electric potential difference caused by the presence of cyanide. This potential is transmitted by a selective electrode and compared to the potential of a reference electrode. In this method, pH>11 should be considered to avoid HCN formation .[63]

3.1.3 Equipment used

For the calibration curve, a Corning model 130 potentiometer was used, with two electrodes, one for reference and the other for cyanide, Orion brand.

III.2 Photocatalysis tests

Photocatalytic cyanide oxidation tests were carried out in a 400 ml beaker. A catalyst concentration of 1.25, 1.0, 0.75, 0.5 and 0.25 g/L of Degussa P-25 titanium dioxide was used.25 g/L of Degussa P-25 titanium dioxide, which is recognised for its high photocatalytic activity, making it the most widely used material in environmental photocatalytic applications; The initial cyanide concentration was 400 ppm, this solution was prepared from NaCN, an anaKtic reagent of 95 % purity, manufactured by Productos Quimicos de Monterrey, a 450 W Regent halogen lamp was used as a light source for 30 minutes with agitation, placing the light source at the top of the beaker for direct irradiation (Figure 3.1). For the determination of cyanide, the cyanide selective ion electrode technique was used with a pH of 11 to avoid the formation of HCN.

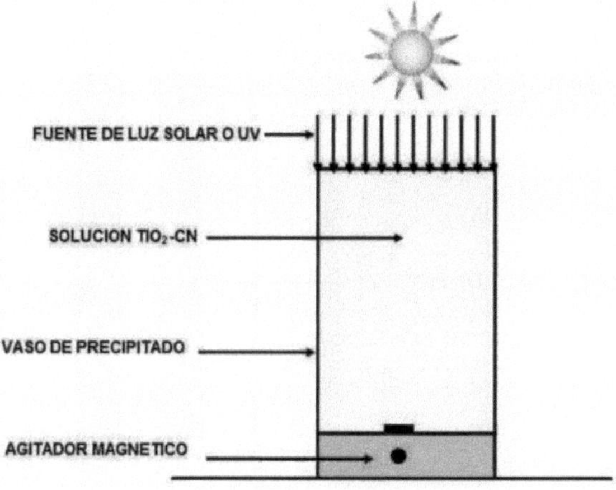

Figure 3.1 Schematic diagram for the photocatalytic oxidation of CN^-.

Table 3.1 summarises the test conditions:

Table 3.1 Photocatalysis test conditions.

Sample	[TiO2] g/L	pH	$[CN]_{initial}$ ppm	Power (W)
1	0.50	11	400	450
2	0.75	11	400	450
3	1.0	11	400	450
4	1.50	11	400	450
5	2.0	11	400	450

III.3 Electrocoagulation tests

3.3.1 Electrocoagulation of TiO2

The electrocoagulation tests for TiO2 were carried out in a 400 ml beaker, using two iron electrodes of 7 cm high by 4 cm wide and with a separation between them of 5 mm. The volume used was 350 ml (Figure 3.2).

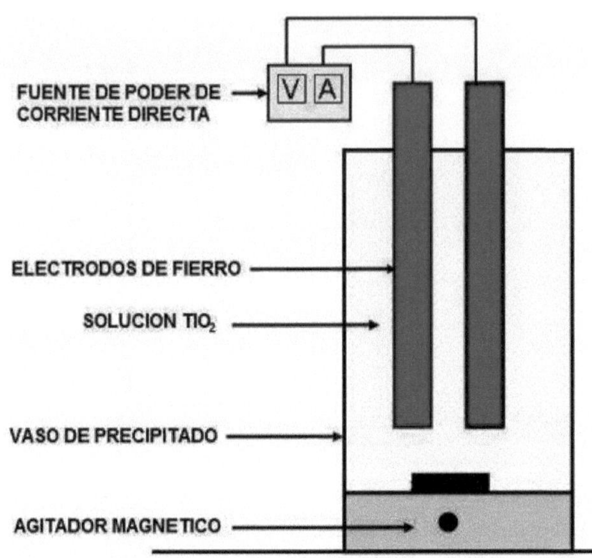

Figure 3.2. Schematic diagram of the electrocoagulation test.

In the electrocoagulation reactor, the iron plates allow the use of O2 and H_2 gases generated from the electrolysis of water to help remove the ferrous and ferric species associated with titanium dioxide. A variable transformer was used to control the voltage and current. The solution was prepared with distilled water and the conductivity was controlled by the addition of 1 g of NaCl per litre of water. The pH was adjusted using a 0.13 M NaOH solution. The solution and solids were separated by filtration through filter paper, the electrocoagulation sludge was dried for 8 hours in an oven at 80 degrees Celsius. In table 3.2 and 3.3 we have the inhaling conditions of operation.

Table 3.2. Inhalation Operation conditions for TiO_2.

Sample (g/L)		pH	Conductivity (MS)	Voltage (volt)	Time (min.)	Current (A)
1	0.50	9	4.10	11.3	30	0.46
2	0.75	9	4.14	10.9	30	0.44
3	1.0	9	4.13	10.62	30	0.45
4	1.50	9	4.10	11.3	30	0.46
5	2.0	9	4.14	10.9	30	0.44
6	2.5	9	4.13	10.62	30	0.45
7	3.0	9	4.14	10.9	30	0.44

To obtain the kinetic data (time-concentration variation), samples were taken every 5 minutes until 30 minutes of operation were completed, 7 samples were taken for each TiO_2 concentration, the number of tests is shown in table 3.3.

Table 3.3 Experiment block for electrocoagulation with TiO_2.

Sample	Time (min)
1	0

2	5
3	10
4	15
5	20
6	25
7	30

To measure the initial and final concentrations of the aqueous samples, atomic absorption spectrometry (Shimadzu equipment model 6701) was used. The solid product was characterised by X-ray diffraction (Phillips X-PERT diffractometer) and scanning electron microscopy (FEI Quanta 2000, Oxford Instruments).

3.3.2 Electrocoagulation As

The electrocoagulation tests for As to determine the number of adsorbed moles were performed as follows: to determine the adsorption and thermodynamic parameters, the tests were performed in a 400 ml beaker, 2 iron electrodes of 6 cm high by 3 cm wide and with a separation between them of 5 miKmeters. The volume used was 350 ml (Figure 3.2). The tests to determine the kinetic parameters were carried out in the 1.3 litre reactor, this reactor has 6 iron plates 15 cm wide by 3 cm wide.

25 cm de alto, figura 3.3.

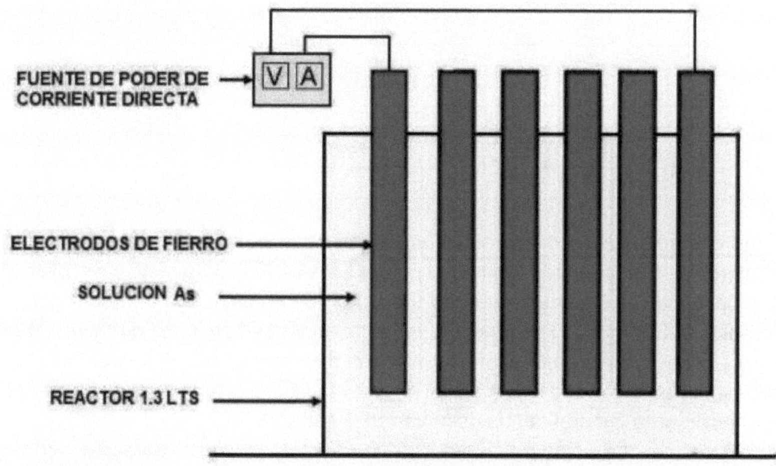

Figure 3.3 Schematic diagram of the 1.3 Litre reactor.

Synthetic arsenic solutions were prepared from 97 % pure anaKtic sodium arsenate (sodium arsenite and arsenic standard can also be used) manufactured by Monterey Chemicals, the pH was adjusted with 0.1 N hydrochloric acid.

Two groups of tests were carried out, the first to determine the number of moles adsorbed, the following concentrations were used: 1, 2, 5, 7, 13, 20 and 30 ppm. The second to determine the kinetic data, with the 5 and 10 ppm samples were taken every 2 minutes until 10 minutes were completed. Tables 3.4 and 3.5 show the initial conditions and the concentrations used in these tests, table 3.6 shows the block of experiments for the second group of tests.

Table 3.4 Operating conditions for As test 1.

Sample (ppm)		pH	Conductivity (RS)	Voltage (volts)	Current (Amperes)	Time (min)
1	1	2.86	4.20	8.5	0.41	5
2	2	2.90	4.10	12.1	0.46	5
3	5	3.05	4.15	10	0.49	5
4	7	2.95	3.90	12.80	0.41	5
5	13	2.90	3.96	10.7	0.46	5
6	20	3.02	4.05	10.25	0.50	5
7	30	4.17	4.12	6.68	0.75	5

Table 3.5. Operation initial conditions for As test 2.

Sample (ppm)	pH	Temperature ° C	Current (Ampere)	Voltage (Volt)	Time (Min)
5	3.55	23	0.65	15.43	30
10	3.26	22	0.63	18.76	30

Table 3.6 Experiment block for the second set of tests.

Sample	Time (min)
1	0
2	2
3	4
4	6
5	8
6	10

For the characterisation of the aqueous samples the plasma spectrometry technique was used, the equipment is Shimadzu model EPA 610C, these tests were carried out in the chemical analysis laboratory of the Corporation Mexicana de Investigacion en Materiales (COMIMSA). The powders were characterised by X-ray diffraction (Phillips X-PERT diffractometer) and scanning electron microscopy (FEI Quanta 2000, Oxford Instruments) at ITS.

III.4 Specific Area Determination using the BET Method

To determine the specific area of the solid product obtained from the electrocoagulation for the recovery of titanium dioxide and arsenic, a Micromeritics model ASP 2020 was used. These tests were carried out at the Department of Metallurgy and Materials of the Colorado School of Mines, located in Golden Colorado, USA.

For this analysis, the samples were treated to remove adsorbed contaminants by degassing the sample for a period of 12 hr and weighed into a sample tube. The samples were then cooled to - 197° C and a known amount of N2 was passed at controlled doses for 12 hr. The amount of nitrogen adsorbed allowed us to obtain the surface area of the open pores of the sample. The operating conditions for this measurement are shown in table 3.7.

Table 3.7 Operating conditions for surface area measurement.

Degassing Time (hrs)	Adsorption Time N2 (hrs)	Bathroom temperature (^{o} C)	Sample size (grams)
12	12	- 197.432	0.680

IV ANALYSIS AND DISCUSSION OF RESULTS

In this chapter a detailed analysis of the experiments carried out at laboratory level for the determination of the adsorption of titanium dioxide and arsenic on the magnetic iron species generated by the electrochemical process of electrocoagulation will be carried out. The first step is to quantify the concentration of cyanide in the solution:

III.5 Calibration curve for cyanide measurement

The calibration curve was obtained from the following experimental data:

Table 4.1 Concentration of CN- in logarithmic form.

mV	Log Concentration CN-
-142	0
-156.70	0.25
-177.68	0.60
200.62	1
-223.56	1.4
-231.12	1.53
-258.28	2

For the measurement of cyanide the following calibration curve was constructed.

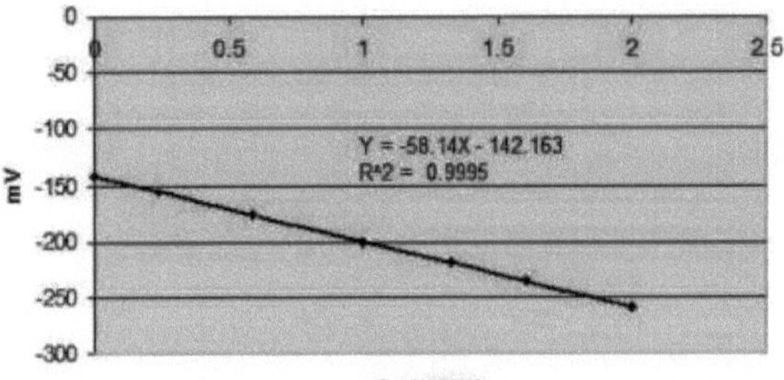

Figure 4.1 Calibration curve for cyanide measurement.

Having the calibration curve, the cyanide concentrations of the in^ial samples and the final samples from the photocatalysis treatment were measured to verify the efficiency of this advanced oxidation process for cyanide removal.

III.6 Photocatalysis test results

The results of cyanide oxidation using photocatalysis technology with TiO2 nanoparticles are shown in table 4.2.

Table 4.2 Cyanide oxidation results.

[TiO2] g/L	$[CN^-]$ initial (ppm)	$[CN^-]$ final (ppm)	% CN elimination -	Time (min)
0.50	400	23	94	30
0.75	400	28	93	30
1.0	400	33	91	30
1.50	400	40	90	30
2.0	400	41	90	30

Table 4.2 shows the results obtained from the oxidation of cyanide, a maximum

recovery of 94 % was obtained with a concentration of titanium dioxide of 0.50 g/L. It is proved that the oxidation of cyanide to cyanate using photocatalysis technology with titanium dioxide is a viable option for eliminating this pollutant from the water, its only limitation being the recovery of the titanium dioxide microelectrodes for reuse in the photocatalytic oxidation process, to remedy this point, the electrocoagulation technology was subsequently used to efficiently recover the TiO_2 microelectrodes.

The same table shows the behaviour of the concentration of TiO2 with the percentage of cyanide elimination, as the concentration of the catalyst increases, the elimination of cyanide decreases, this is due to the fact that an optical shielding occurs as the concentration of the catalyst increases, thus decreasing the generation of OH- radicals for the oxidation of cyanide, which was also observed by other researchers .[60]

III.7 TiO_2 Electrocoagulation Test Results

4.3.1 Results of aqueous samples

Table 4.3 shows the results obtained from the aqueous TiO2 samples using the atomic absorption spectrometry technique, these results were used for the adsorption study.

Table 4.3 Results of aqueous TiO2 samples (adsorption study).

Sample	$[TiO_2]_{initial}$ g/L	$[TiO2]_{final}$ g/L	Recovered
1	0.5	0.0085	98.3
2	0.75	0.0225	97
3	1	0.034	96.6
4	1.5	0.052	96.53
5	2	0.083	95.85
6	2.5	0.095	96.2
7	3	0.15	95.0

With the results obtained in the table above, the high efficiency of TiO2 recovery using the electrochemical process of electrocoagulation is proved, thus solving the weak point of this technology, which is the recovery of titanium dioxide and its subsequent reuse in the photocatalytic oxidation of cyanide.

The higher recovery values also correspond to the higher amount of current density applied due to the higher production of Fe^{+3} thus improving the recovery due to the higher probability of interaction between the generated EC species and titanium dioxide, The decrease in recovery efficiency is due to the fact that in acidic solutions, the oxidation of ferrous iron (Fe^{+2}) to ferric iron (Fe^{+3}) decreases and therefore the recovery of titanium dioxide decreases; at higher pH the tendency is in favour of the oxidation of Fe^{+2} to Fe^{+3} thus improving the electrocoagulation process for titanium dioxide recovery.

The results in table 4.3 were also used to perform the adsorption calculations and subsequently to obtain the thermodynamic parameters of the adsorption of titanium dioxide on EC-generated species.

The results for the kinetic study are shown in the following tables. In these experiments, a sample was taken every 5 minutes until 30 minutes were completed to obtain the titanium dioxide concentration-time variation for the kinetic study.

Table 4.4 Results obtained for sample 1 TiO2 (kinetic study).

Time (min)	$[TiO_2]_{initial}$ (g/L)	$[TiO2]_{final}$ (g/L)	Recovered
O	0.5	0.5	0
5	0.5	0.39	22
10	0.5	0.275	45
15	0.5	0.18	64
20	0.5	0.07	86
25	0.5	0.03	94
30	0.5	0.01	98

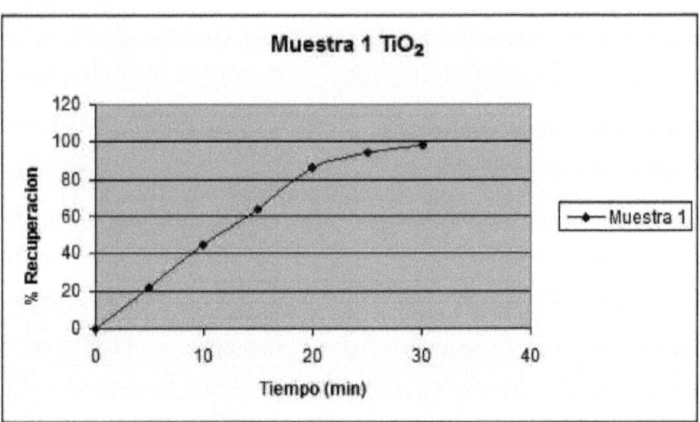

Figure 4.2 Graph of TIO2 recovery - time for sample 1.

From the results shown in table 4.4, we have a titanium dioxide recovery of 98 %, which shows that electrocoagulation is a viable alternative to recover titanium dioxide and subsequently reuse it in the photocatalysis process for cyanide removal.

Table 4.5 Results obtained for sample 2 TIO2 (kinetic study).

Weather	initial $[TiO2]$ (g/L)	$[TiO2]_{final}$ (g/L)	Recovered
O	0.75	0.75	0
5	0.75	0.62	17
10	0.75	0.46	37
15	0.75	0.32	57
20	0.75	0.22	71
25	0.75	0.11	85
30	0.75	0.06	92

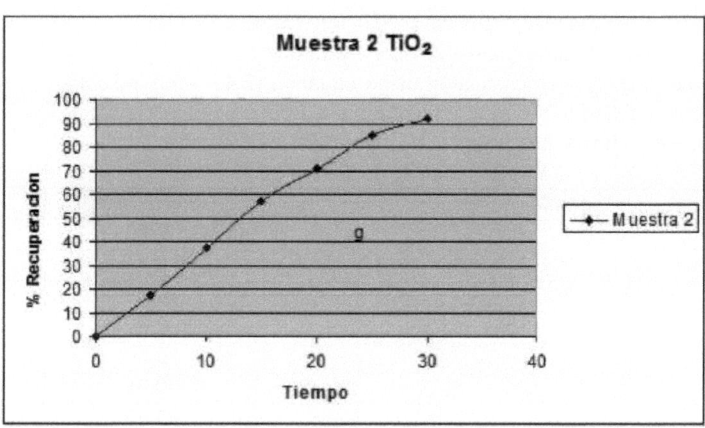

Figure 4.3 Plot of TłO2 recovery versus time for sample 2.

The result obtained for sample 2 is 92 %, this decrease in the recovery is due to the variation of the current supplied, being in this case lower than sample 1 (0.46 Amp. for sample 1 and 0.44 Amp. sample 2).

Table 4.6 Results obtained for sample 3 TłO2 (kinetic study).

Weather	initial [TiO2] (g/L)	[TiO2]final (g/L)	Recovered
O	1	1	0
5	1	0.76	24
10	1	0.52	48
15	1	0.34	66
20	1	0.22	78
25	1	0.15	85
30	1	0.04	96

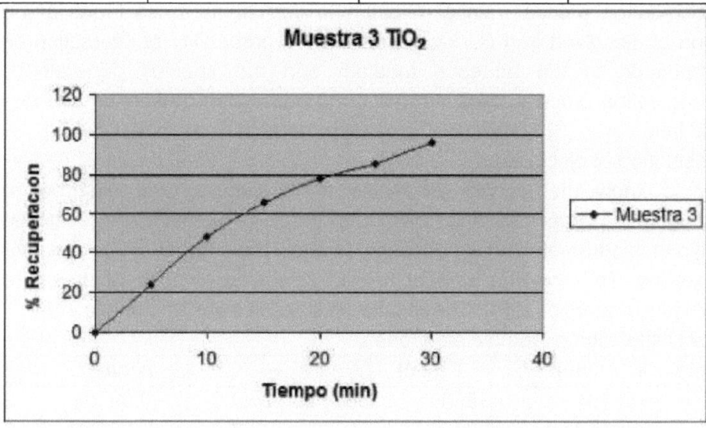

Figure 4.4 Graph of TiO2 recovery versus time for sample 3.

This sample has a recovery of 96 %, this is due to the difference in the current density applied to the electrodes (sample 3, 0.5 Amp).

4.3.2 Calculation of current density

The current density was calculated using equation (2.47), and indicates the ratio of the current supplied to the total area of the electrodes used in the electrocoagulation tests, table 4.7 shows the current used in the tests performed.

Table 4.7 Current applied to each sample.

Sample	Current (Amp)
1	0.46
2	0.44
3	0.45
4	0.47
5	0.46
6	0.44
7	0.45

Table 4.8 Current density.

Sample	Current (Amp)	Area (cm)^2	D(Amp/cm)^2
1	0.46	28	0.0164
2	0.44	28	0.0157
3	0.45	28	0.0161
4	0.47	28	0.0168
5	0.46	28	0.0164
6	0.44	28	0.0157
7	0.45	28	0.0161

The current density is one of the most important parameters in the EC tests, since the amount of species generated in the EC, such as magnetite, lepidocrocite, goethite, depends on it, thus improving the efficiency of the recovery or elimination of the electrocoagulation process, since at optimum current densities there is a greater production of dissolved iron ($Fe+^{3}$), increasing the probability of interaction between the compounds in the aqueous medium and the species generated in the electrocoagulation process, thus improving the adsorption of titanium dioxide on the dissolved iron.

4.3.3 Dissolution of electrodes

In order to know the number of moles of titanium dioxide adsorbed on the electrocoagulation species, and with this data to subsequently calculate the thermodynamic parameters, it is necessary to know the amount of iron dissolved from the electrodes. To know the amount of iron dissolved in each treatment, we use Faraday's law, equation (2.48). The results obtained are shown in table 4.9.

Table 4.9 Dissolution of sample electrodes.

Sample	D(A/cm)^2	T (sec)	M	N	F	W(g/cm$^{2)}$	grams.
1	0.0164	1800	56	3	96500	0.00191	0.107
2	0.0157	1800	56	3	96500	0.00182	0.102
3	0.0161	1800	56	3	96500	0.00187	0.104
4	0.0168	1800	56	3	96500	0.00195	0.109

5	0.0164	1800	56	3	96500	0.00191	0.107
6	0.0157	1800	56	3	96500	0.00182	0.102
7	0.0161	1800	56	3	96500	0.00186	0.104

Knowing the amount of dissolved iron for each experiment performed, it is possible to calculate different adsorption data such as the number of adsorbed moles (N), maximum adsorption number (Nmax), fraction covered and separation factor R_L.

4.3.4 Energy consumption

The energy consumption of an electrocoagulation reactor can be calculated with equation (2.49). The results obtained are shown in table 4.10.

Table 4.10 Energy consumption for each sample.

Sample	I (Amp)	V (volt)	t (hr)	E (KWhr)
1	0.46	9.3	0.5	0.002139
2	0.44	10.12	0.5	0.0022264
3	0.45	9.5	0.5	0.0021375
4	0.47	9,50	0.5	0.00210325
5	0.46	8.95	0.5	0.0022195
6	0.44	9.65	0.5	0.002244
7	0.45	10.2	0.5	0.002223

According to the data shown in table 4.10, the total energy consumption is 0.015 kWhr, which is one of the advantages of using the EC process, its low energy consumption thus ensuring low operating costs.

4.3.5 Cost of treatment

Energy cost: 3.32 KWhr

C_{EI} = Energy cost per treatment in sample 1

C_{E2} = Energy cost per treatment in sample 2

C_{E3} = Cost of energy per treatment in sample 3

Table 4.11 Treatment cost for each sample taken.

Sample	Energy costs	E(KWhr)	C_E ($)
1	3.32	0.00214	0.00710
2	3.32	0.00222	0.00739
3	3.32	0.00213	0.00709
4	3.32	0.00210	0.00698
5	3.32	0.00221	0.00736
6	3.32	0.00224	0.00745
7	3.32	0.00222	0.00738

The total cost of the EC treatments is $ 0.018/litre of treated water, which demonstrates the low costs of EC operation. One of the most important parameters affecting the application of any method to treat contaminated water or to recover reagents such as titanium dioxide is the cost. The operating cost includes material, mainly electrodes and electricity costs, as well as maintenance, disposal and sludge drying costs. In table 4.11 we can see the costs used for each sample, which shows one of the main advantages of electrocoagulation, its low operating cost compared to other chemical or biological systems.

4.3.6 Number of moles adsorbed per gram of adsorbent

The number of moles adsorbed per gram of adsorbent was calculated by equation (2.65). The results obtained for sample 1 are shown in table 4.12.

Table 4.12 Moles adsorbed per gram of adsorbent.

Sample	Co(mmol/L)	C(mmol/L)	mc(grs)	N(mmolTio2/gFe)	C/N
1	6.259	0.125	0.160	13.405	0.00933
2	9.389	0.281	0.153	20.807	0.0135
3	12.519	0.475	0.157	26.902	0.0177
4	18.778	0.651	0.167	37.962	0.0171
5	25.037	1.039	0.164	51.326	0.0202
6	31.297	1.189	0.157	67.254	0.0177
7	37.556	1.878	0.160	77.966	0.0241

With the values obtained for N, the amount of moles of titanium dioxide adsorbed per gram of adsorbent is obtained. This value is calculated by the amount of dissolved iron, so the behaviour is the same as that of the iron solution and is a function of the amount of current supplied.

The difference in N values implies different pore sizes of the adsorbent, due to the complex distribution of the total pore volume in the iron hydroxides.

Using equation (2.67) (linearised Langmuir isotherm equation) and plotting C/N with C (graph 4.4), we obtain the following linear regression data:

Table 4.13 Results obtained from the linear regression.

m (slope)	0.0103
and (ordered)	0.361
R	0.994
R2	0.988

If the system follows the behaviour described by the Langmuir isotherm, the graph of the C/N ratio as a function of the equilibrium concentration C should give a straight line of slope 1/Nmax and ordinate to the origin 1/KNmax. The results obtained are:

Table 4.14 Obtaining Nmax and K.

Nmax (mmolTiO2)/gFe)	96.7
K $(Lmmol)^{-1}$	35.12

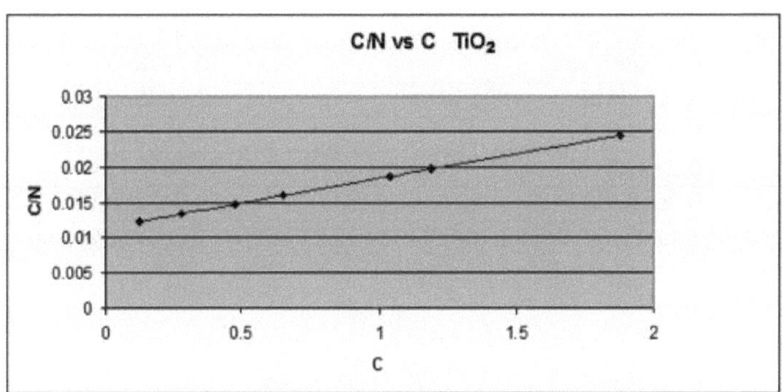

Figure 4.5 C/N vs C graph to calculate Nmax.

C= Final concentration.

N= Number of moles adsorbed.

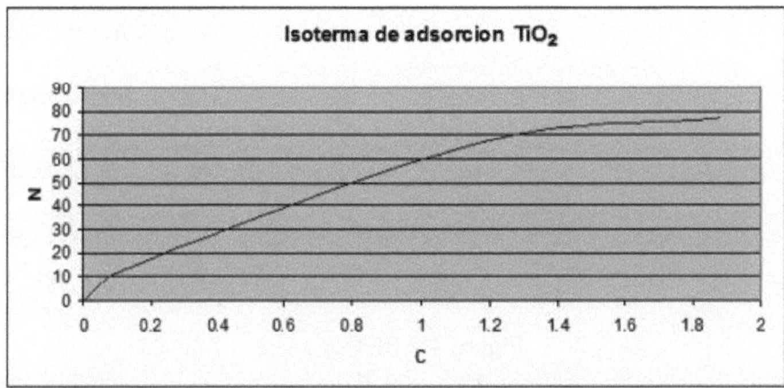

Figure 4.6 Adsorption isotherm of TIO2 on iron hydroxides.

C= Final concentration.

N= Number of moles adsorbed).

Figure 4.5 shows the line that adjusts the data, by the obtained value of the correlation coefficient (0.994) it is verified that the system follows the behaviour of the Langmuir isotherm. Figure 4.6 shows the data of the experimental adsorption isotherm, in which we can appreciate the behaviour of the adsorption system, where we can first observe the formation of a multilayer, followed by the formation of a monolayer at higher concentrations, and then the adsorption remains constant reaching the maximum number of adsorption.

The value of Nmax is the maximum adsorption capacity of the titanium dioxide corresponding to the complete coating of the monolayer on the iron surface, with this value the fraction covered is calculated for each sample.

4.3.7 Specific area

The BET method of nitrogen adsorption was used to determine the specific area, which assumes that the nitrogen gas when liquefied and adsorbed on solid surfaces will fill all the available clean surface area forming multiple layers. The results obtained for nitrogen adsorption are shown in the table below:

Table 4.15 Nitrogen adsorption.

Relative Pressure	Amount Adsorbed

Po/P	(cm^3/g)
0.05688504	10.7978951
0.1121802	12.0635681
0.17515116	13.4282578
0.23775899	14.7837413
0.30047937	16.0437507

In the following figure we have the graph of the BET isotherm corresponding to the nitrogen adsorption.

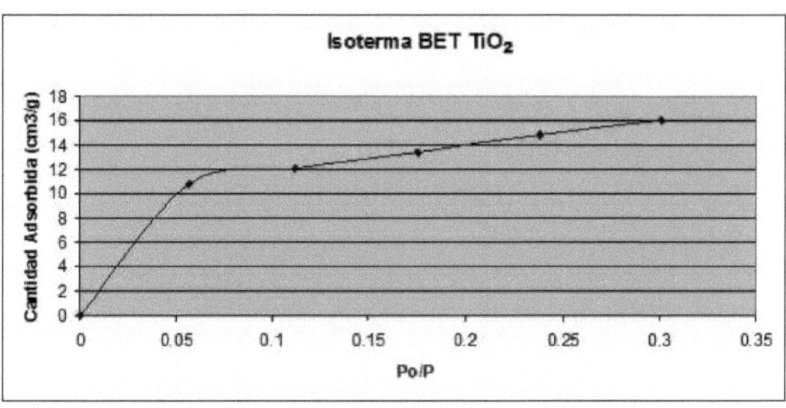

Figure 4.7 BET chart.

To calculate the specific surface area, the data in table 4.16 is plotted using the linearised BET equation (equation 2.68), then linear regression is applied to the data and using equations (2.72) and (2.73) the specific surface area is calculated.

Table 4.16 Data to calculate specific area.

Relative Pressure Po/P	1/[Q(Po/P - 1)]
0.05688504	0.00558592
0.1121802	0.01047407
0.17515116	0.01581317
0.23775899	0.02109892
0.30047937	0.02677369

Table 4.17 Data obtained from the linear regression.

Pending	0.0864
Ordered to the origin	0.000685
R	0.9999
R2	0.9998

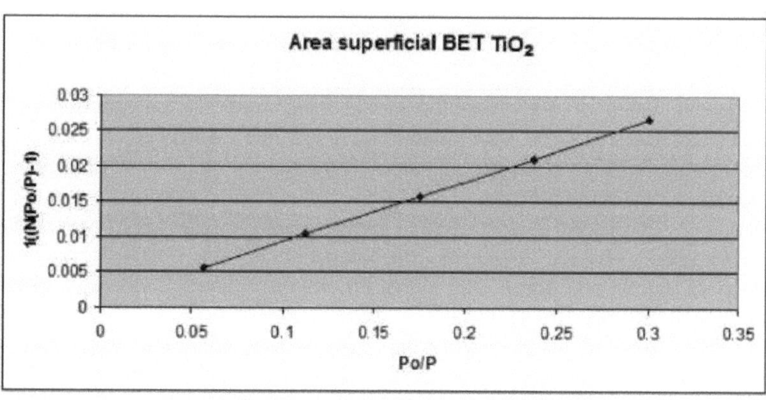

Figure 4.8 Line to obtain the surface area with the BET method.

The specific area value obtained is 190 m^2 /g. This value is within the usual range for adsorbents consisting of small and porous particles, between 10 and 1 000 m^2 /g. The value of the specific area also determines the adsorption capacity of the adsorbent being used, in this case the species generated from EC, this adsorption capacity is verified with the values obtained in the recovery of titanium dioxide (98%).

4.3.8 Fraction of covered area

According to the results obtained for N and N_{max}, the fraction of surface area covered *Q* is calculated by the equation $Q = N/N_{max}$. From table 4.18 it can be seen that the fraction of coverage approaches 1 as the concentration increases.

Table 4.18 Obtaining *Q*.

N (mmolTio2/gFe)	N_{max} (mmolTiO2)/gFe)	Q
13.405	96.7	0.139
20.807	96.7	0.217
26.902	96.7	0.280
37.962	96.7	0.395
51.326	96.7	0.535
67.254	96.7	0.700
77.966	96.7	0.812

The value of *Q* varies between 0 and 1, having a minimum fraction covered value of 13.9% corresponding to the obtained N value of 13.405 mmolTiO2)/gFe and the maximum fraction obtained of 81.2% with the N value of 77.966 mmolTiO2)/gFe, therefore the results obtained are within the expected range.

4.3.9 Calculation of thermodynamic parameters

Applying equations (2.74), (2.75) and (2.76) we obtain the following results for free energy, enta^a and entrop^a for the adsorption process using 7 concentrations.

Table 4.19 Thermodynamic parameters.

	Kcal/mol	Kj/mol
AG	-8.146	-39.0184
AH	-4.145	-20.397

	Kcal/molK	Kj/molK
AS	0.0305	0.1473

The negative value of AG presented in the table confirms the feasibility of the adsorption process and the spontaneous nature of the adsorption of TiO2 on the species generated from electrocoagulation. The negative value of AH indicates the exothermic nature of the process.

The value of adsorption HA according to the literature corresponds to typical values for physisorption (Typical values of adsorption enthalpy for physisorption are -20 kj/mol to -80 kj/mol and about -200 kj/mol for chemisorption[57]), the value obtained being -29.0184 kJ/mol.

The positive value of AS shows the increase of randomness at the solute-solid interface during the adsorption process, suggesting that the adsorbed titanium dioxide replaces some water molecules from the solution previously adsorbed on the adsorbent surface.

4.3.10 RL factor

The results obtained for the separation factor are shown in the table below:

Table 4.20 Separation factor.

Sample	Co (mmol/L)	B (mmol/g)	RL
1	6.259	35.12	0.000657
2	9.389	35.12	0.000438
3	12.518	35.12	0.000328
4	18.778	35.12	0.000219
5	25.037	35.12	0.000164
6	31.297	35.12	0.000131
7	37.556	35.12	0.000109

According to the obtained results, the separation factor is in the range of 0 and 1, which represents that the adsorption system of TiO2 and electrocoagulation-generated species such as magnetite, goethite and lepidocrocite is favourable.

4.3.11 Calculation of kinetic parameters

4.3.11.1 Calculation of kinetic and adsorption constant

4.3.11.1.1 Sample 1 TiO2

Table 4.21 shows the variation of time versus TiO2 concentration data for the calculation of the kinetic parameters.

Table 4.21 Time-concentration data sample 1 (kinetic study).

Time (min.)	Concentration TiO2 (g/L)
0	0.5
5	0.39
10	0.275
15	0.18
20	0.07
25	0.03
30	0.01

Exponential regression was applied to the data in table 4.21 to obtain the following data shown in table 4.22.

Table 4.22 Statistical parameters obtained from the regression.

R	0.9918
R^2	0.918
b	530
m	0.00805

To find the kinetic reaction constant k and the adsorption rate constant K, the linearised equation of the Langmuir-Hinshelwood model, equation (2.78) is used using the experimental data shown in table 4.23:

Table 4.23 Experimental data for using the Langmuir-Hinshelwood equation.

-dt/dC	1/C
23.52941176	2
30.16591252	2.56410256
42.78074866	3.63636364
65.35947712	5.55555556
168.0672269	14.2857143
392.1568627	33.3333333
1176.470588	100

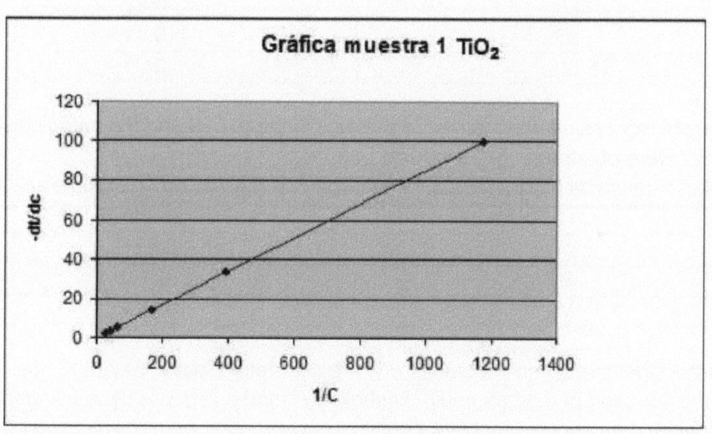

Figure 4.9 Plot of experimental data -dt/dC versus 1/C.

Linear regression is applied to the data in table 4.23 and the following values are obtained:

Table 4.24 Statistical parameters obtained from the regression.

R	1
R2	1
b	6.45E-9
m	11.76

Applying equations (2.82) and (2.83) gives the kinetic constants shown in table 4.25.

Table 4.25 Kinetic parameters sample 1.

Kinetic constant k	15.50 E4 g L^{-1} min^{-1}
Adsorption constant K	5.5 E-7 L g^{-1}

From table 4.26 we have that the kinetic constant (k) is larger than the adsorption constant (K), which means that the controlling mechanism for the TiO2 recovery process with electrocoagulation is the rate of adsorption of titanium dioxide (stage 2 of the adsorption process) on the surface of the species generated from electrocoagulation rather than the rate of conversion of the same (stage 3 of the adsorption process).

4.3.11.1.2 Sample 2 TiO2

Table 4.26 shows the variation of time versus TiO2 concentration data for the kinetic study for sample 2 corresponding to a concentration of 0.75 g/L.

Table 4.26 Time-concentration data sample 2 (Kinetic study).

Time (min.)	Concentration TiO2 (g/L)
0	0.75
5	0.62
10	0.46
15	0.32
20	0.22
25	0.11
30	0.06

Exponential regression was applied to the data in table 4.26 and the data shown in table 4.27 were obtained.

Table 4.27 Statistical parameters obtained from the regression on sample 2.

R	0.9957
R2	0.9913
B	0.793
M	0.0643

To find the kinetic reaction constant k and the adsorption rate constant K, the linearised equation of the Langmuir-Hinshelwood model (2.78) is used using the experimental data shown in table 4.29:

Table 4.28 Experimental data for using the Langmuir-Hinshelwood equation.

-dt/dc	1/C
20.7361327	1.33333333
25.0840315	1.61290323
33.808912	2.17391304
48.600311	3.125
70.6913615	4.54545455
141.382723	9.09090909
259.201659	16.6666667

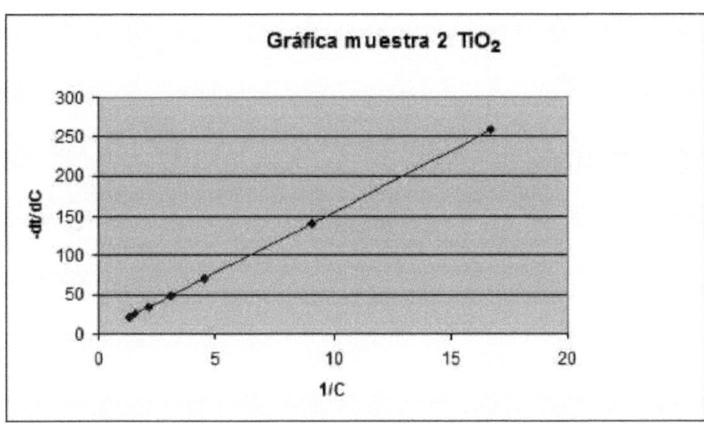

Figure 4.10 Plot of experimental data -dt/dC versus 1/C.

Linear regression was applied to the data in table 4.28 and the values shown in table 4.29 were obtained.

Table 4.29 Statistical parameters obtained from regression sample 2.

R	1
R2	1
b	9.44 E-8
m	15.55

Applying equations (2.82) and (2.83), the kinetic constants shown in table 4.30 are obtained.

Table 4.30 Kinetic parameters sample 2.

Kinetic constant k	10.59 E3 mg L^{-1} min^{-1}
Adsorption constant K	6E-6 L mg^{-1}

From table 4.30 we have that the kinetic constant (k) is larger than the adsorption constant (K), which means that the controlling mechanism for the TiO2 recovery process with electrocoagulation is the rate of adsorption of the titanium dioxide (stage 2 of the adsorption process) on the surface of the generated electrocoagulation species, displacing the rate of reaction of the same (stage 3 of the adsorption process).

4.3.11.1.3 Sample 3 TiO2

Table 4.31 shows the variation of the time versus TiO2 concentration data for the kinetic study for the titanium dioxide sample corresponding to 1 g/L.

Table 4.31 Time-concentration data sample 3 (Kinetic study).

Time (min.)	Concentration TiO2 (g/L)
0	1
5	0.76
10	0.52
15	0.34

20	0.22
25	0.15
30	0.04

Exponential regression was applied to the data in table 4.31 and the data shown in table 4.32 were obtained:

Table 4.32 Statistical parameters obtained from the regression Sample 3.

R	0.9977
R^2	0.9953
B	1.053
M	0.0748

To find the kinetic reaction constant k and the adsorption rate constant K, the linearised equation of the Langmuir-Hinshelwood model (2.78) is used using the experimental data shown in table 4.33:

Table 4.33 Experimental data for using the Langmuir-Hinshelwood equation sample 3.

-dt/dc	1/C
15.5520995	1
20.4632889	1.31578947
29.9078837	1.92307692
45.7414692	2.94117647
70.6913615	4.54545455
103.680664	6.66666667
388.802488	25

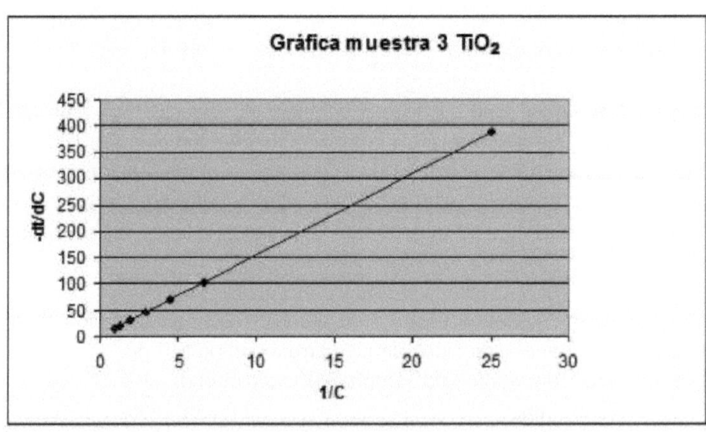

Figure 4.11 Plot of experimental data -dt/dC versus 1/C.

Linear regression is applied to the data in table 4.33 and the values shown in table 4.34 are obtained:

Table 4.34 Statistical parameters obtained from the regression sample 3.

R	1
R^2	1
B	4.81 E-9
M	15.55

Applying equations (2.82) and (2.83), the kinetic constants are obtained.

Table 4.35 Kinetic parameters sample 3.

Kinetic constant k	20.8E4 mg L^{-1} min^{-1}
Adsorption constant K	3.09 E-7 L mg^{-1}

In table 4.35 we can verify that the kinetic constant (k) is higher than the adsorption constant (K), which means that the controlling mechanism for the TiO2 recovery process with electrocoagulation is the rate of adsorption of the titanium dioxide (stage 2 of the adsorption process) on the surface of the species generated from electrocoagulation, displacing the rate of conversion of the same (stage 3 of the adsorption process).

4.3.11.2 Calculation of reaction order

The most widely used model to describe the kinetics of the photocatalytic process is the Langmuir-Hinshenlwood model (Cassano, 1999). To know the order of the reaction, equation (2.77) is used, which is the expression of the first order kinetics, its integrated form is (($-\ln(C_f/C_o)=kt$). When evaluating the kinetic data by means of this model, the following results were obtained.

4.3.11.2.1 Sample 1

To calculate the order of the reaction, $\ln(C_f/C_o)$ is calculated, where Cf is the final concentration and C_o is the initial concentration, then the data of $\ln(C_f/C_o)$ and time are linearly regressed to find the correlation coefficient and thus verify the order of reaction 1.

Table 4.36 Data to determine the reaction order.

Weather	Cf	Co	Cf/Co	$-\ln(C_f/C_o)$
0	0.5	0.5	1	0
5	0.39	0.5	0.78	0.320
10	0.275	0.5	0.75	0.734
15	0.18	0.5	0.36	0.950
20	0.07	0.5	0.14	1.080
25	0.03	0.5	0.06	1.327
30	0.01	0.5	0.02	1.623

The data obtained from the linear regression are shown in the table below:

Table 4.37 Results obtained from the linear regression.

Pending	0.0516
Ordered to the origin	0.0875
R	0.990
R^2	0.980

The resulting linear regression plot is shown in figure 4.12.

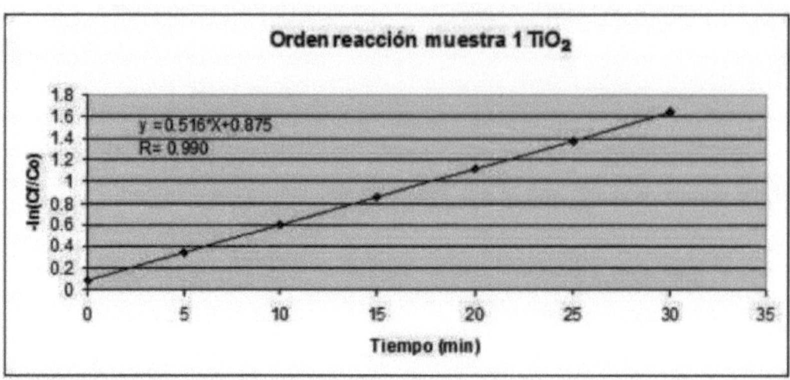

Figure 4.12 Evaluation of first-order kinetics.

These results verify that the data obtained fit an order 1 model.

4.3.10.2.2 Sample 2

To calculate the order of the reaction, ln(cf/co) is calculated, where Cf is the final concentration and co is the initial concentration, then the data of ln(cf/co) and time are linearly regressed to find the correlation coefficient and thus verify the order of reaction 1.

Table 4.38 Data to determine the reaction order.

Time (min)	Cf (g/L)	Co(g/L)	Cf/Co	-ln(cf/co)
0	0.75	0.75	1	0
5	0.62	0.75	0.8267	0.1904
10	0.46	0.75	0.6133	0.4888
15	0.32	0.75	0.4267	0.8518
20	0.22	0.75	0.2933	1.2264
25	0.11	0.75	0.1467	1.9196
30	0.06	0.75	0.080	2.5257

The data obtained from the linear regression are shown in the table below:

Table 4.39 Results obtained from the linear regression.

Pending	0.0679
Ordered to the origin	0.0725
R	0.950
R2	0.90

The resulting linear regression plot is shown in Figure 4.13.

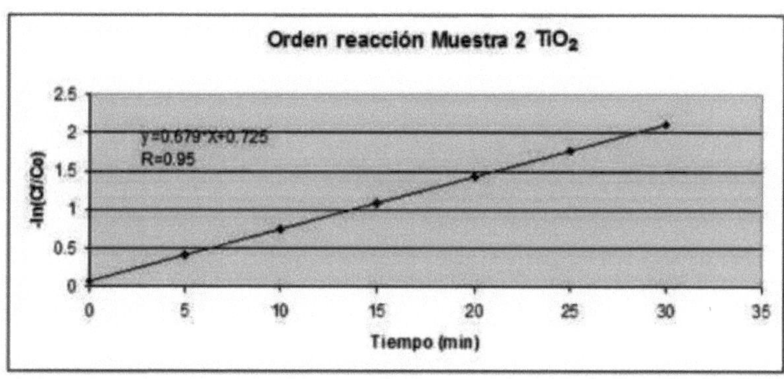

Figure 4.13 Evaluation of first-order kinetics.

These results verify that the data obtained fit an order 1 model.

4.3.10.2.3 Sample 3

In order to know the order of the reaction, ln(cf/co) is calculated, where Cf is the final concentration and co is the initial concentration, then the data together with the time is applied linear regression to know the correlation coefficient and in this way verify the order of the reaction 1.

Table 4.40 Data for determining the order of reaction.

Time (min)	cf(g/L)	Co(g/L)	Cf/Co	-ln(cf/co)
0	1	1	1	0
5	0.76	1	0.76	0.27443685
10	0.52	1	0.52	0.65392647
15	0.34	1	0.34	1.07880966
20	0.22	1	0.22	1.51412773
25	0.15	1	0.15	1.89711998
30	0.04	1	0.04	3.21887582

The data obtained from the linear regression are shown in the table below:

Table 4.41 Results obtained from the linear regression.

Pending	0.0781
Ordered to the origin	0.1285
R	0.930
R2	0.883

The resulting linear regression plot is shown in figure 4.14.

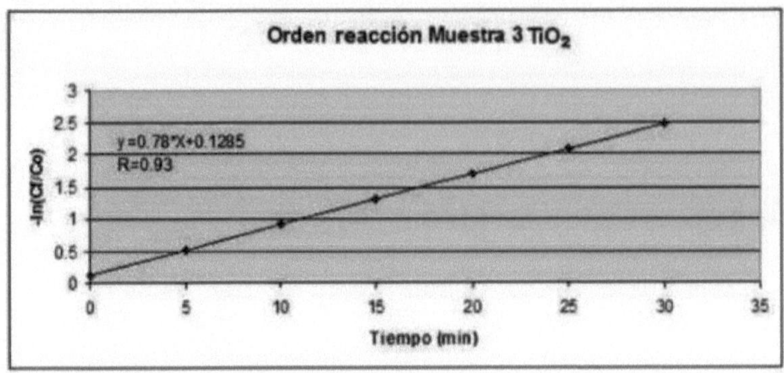

Figure 4.14 Evaluation of first-order kinetics.

These results verify that the data obtained fit an order 1 model.

4.4 Results obtained from electrocoagulation of As

4.4.1 Results obtained from aqueous samples

The results obtained for arsenic removal from the EC liquid samples for the first set of tests for adsorption calculations are shown in table 4.42.

Table 4.42 Results for the first group of tests (adsorption study).

Sample	Initial [As] (ppm)	$[As]_{Final}$ (ppm)	Recovered
1	1	0.04	96
2	2	0.04	98
3	5	0.1	98
4	7	0.35	95
5	13	0.39	97
6	20	0.4	98
7	30	0.3	99

The following graphs show the results obtained in table 4.42.

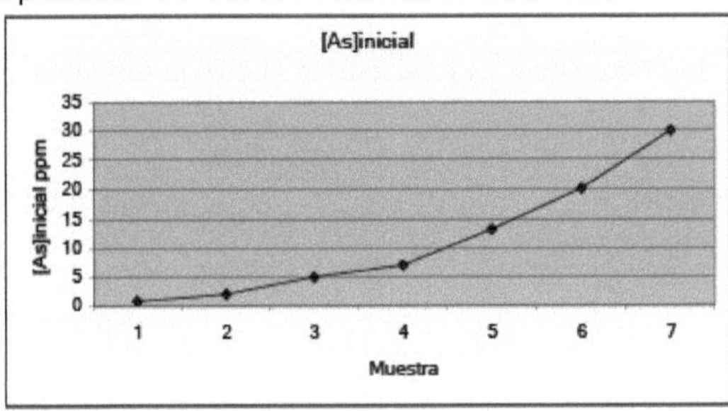

Figure 4.15 Initial As concentration (ppm) and sample number.

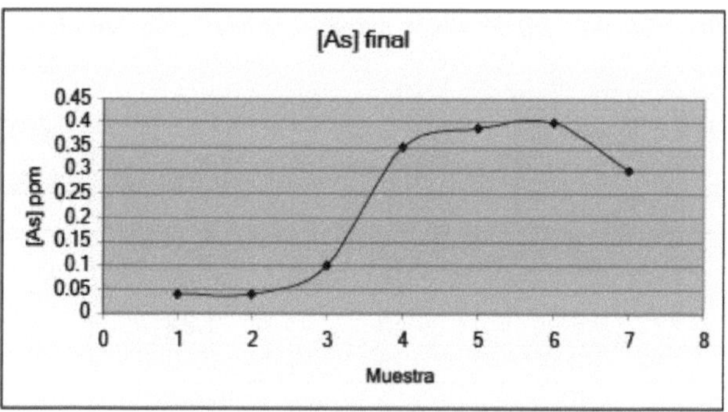

Figure 4.16 Final As concentration (ppm).

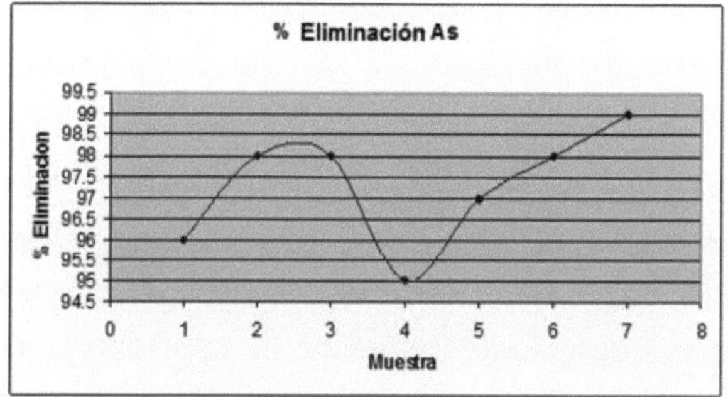

Figure 4.17 % removal obtained for each As sample.

With the results obtained for the first group of tests (to perform the thermodynamic analysis) the efficiency of the electrocoagulation process is verified, having a maximum removal of 99 % of arsenic. This behaviour is consistent with the application of current density for each sample, being the arsenic removal higher for sample 7, which has the highest current applied. The increase in arsenic removal with increasing current density is due to the increased generation of iron hydroxides as more current passes through the electrodes, increasing the probability of interaction between the EC-generated species and arsenic in the water. The increase in arsenic removal can also be explained by the intensity of the electric field (this intensity varies according to the separation of the electrodes, taking into account the obtained removals the separation is optimal) if the intensity of the electric field increases, the electrostatic attraction also increases and the metal removal decreases, conversely, when the intensity decreases, the cathodic attraction decreases, thus increasing the metal removal.

Furthermore, this is one of the advantages of this process over traditional heavy metal removal methods such as chemical precipitation, which does not completely remove arsenic, making it a viable alternative for the removal of this pollutant.

The results obtained for the second group of tests (to obtain the kinetic data by varying the concentration with time) are shown in tables 4.43 (5 ppm) and 4.44 (10 ppm), in these tests a sample was taken every 2 minutes until completing 10, for each time the accumulated percentage is shown, until reaching the maximum elimination value obtained (99 %) which in this case was obtained after 10 minutes of operation.

Table 4.43 Test group 2 results sample 5 ppm (kinetic study).

Sample 1	Time (min)	$[As]_{initial}$ (ppm)	$[As]_{final}$ (ppm)	% Eliminated As
1	0	5	5	0
2	2	5	3.8	24
3	4	5	2.6	48
4	6	5	1.85	63
5	8	5	0.95	81
6	10	5	0.045	99.1

Figures 4.18 and 4.19 show the data obtained in graphical form.

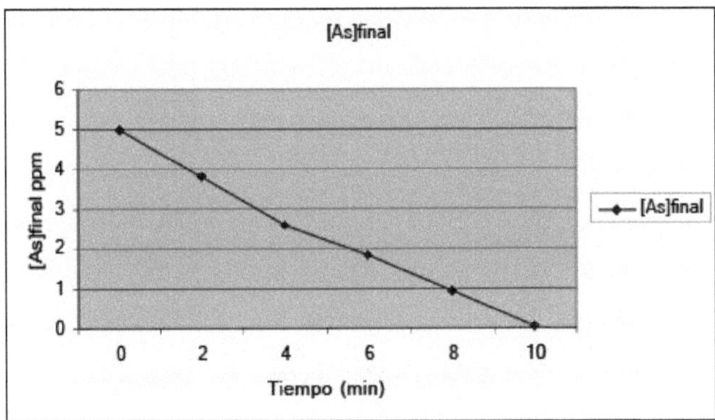

Figure 4.18 Initial As concentration, sample 5 ppm.

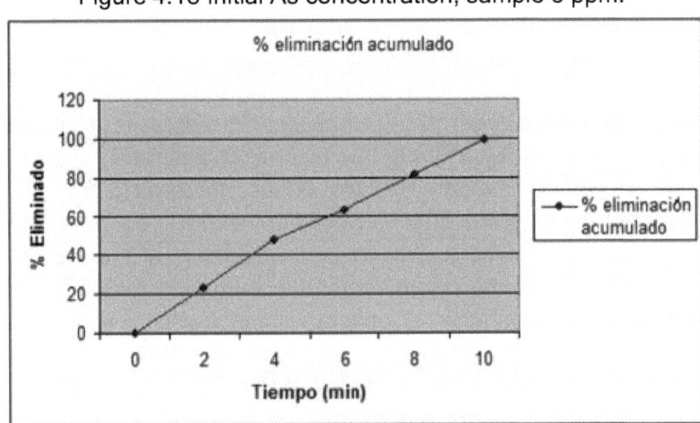

Figure 4.19 % removal as a function of time for the 5 ppm sample.

Table 4.44 shows the results obtained for sample 2, corresponding to a concentration of 10 ppm As.

Table 4.44 Results of test group 2 sample 10 ppm (kinetic study).

Sample 1	Time (min)	Initial [As] (ppm)	$[As]_{final}$ (ppm)	% Eliminated
1	0	10	10	0

2	2	10	7.9	21
3	4	10	5.7	43
4	6	10	3.4	66
5	8	10	1.6	84
6	10	10	0.05	99.5

The data obtained for sample 2 are shown graphically in figures 4.20 and 4.21.

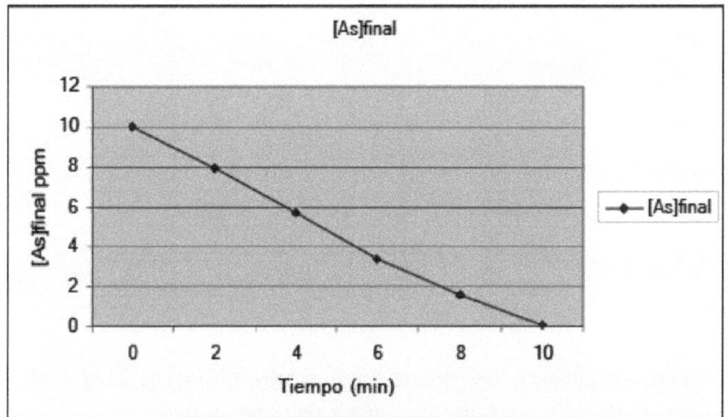

Figure 4.20 Final sample concentration 10 ppm As.

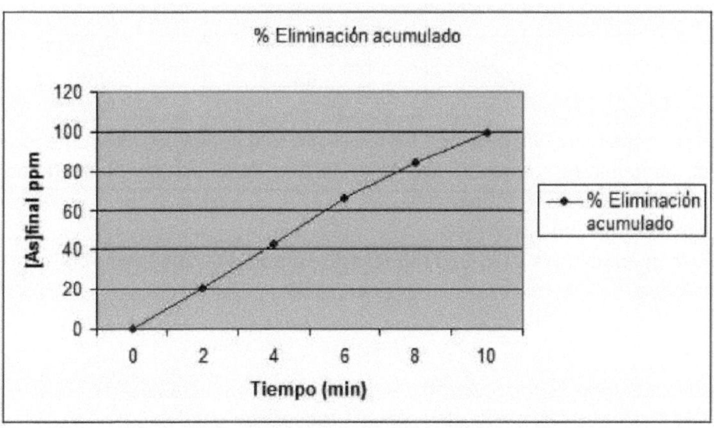

Figure 4.21 % As elimination, sample 10 ppm.

For this group of tests (to carry out the kinetic analysis varying concentration-time, tables 4.43 and 4.44), we have a % of elimination for sample 1 (5 ppm) of 99.1 % and for sample 2 (10 ppm) of 99.55 %. The eliminations obtained have the same behaviour as the first group of tests (table 4.42), the greater amount of current density applied corresponds to the higher % recovery, this is due to the greater generation of iron hydroxides to adsorb the arsenic and thus eliminate this pollutant from the water. These results confirm the high efficiency of this electrochemical (EC) process over conventional heavy metal removal techniques such as chemical precipitation, making it a viable, safe and economical alternative for removing arsenic from contaminated water. The concentrations used are those recommended for the thermodynamic and kinetic study, to remove arsenic from the final samples and bring them within the norm for discharge to effluents, it is recommended to re-treat these samples with EC for 15 minutes.

4.4.2 Calculation of current density

The current density is calculated using equation (2.47). The results are shown in table 4.45. The data are for the first set of tests.

Table 4.45 Current density group 1 (adsorption study).

Sample	Area (cm)^2	Current (Amps)	Current density (Amp/cm)^2
1	36	0.41	0.01138889
2	36	0.46	0.01277778
3	36	0.49	0.01361111
4	36	0.41	0.01138889
5	36	0.46	0.01277778
6	36	0.51	0.01388889
7	36	0.75	0.02083333

For the second group of tests the results obtained are shown in table 4.46.

Table 4.46 Current density second set of tests (Kinetic study).

Sample	Area (cm)^2	Current (Amps)	Current density (Amp/cm)^2
1	375	0.65	0.00173
2	375	0.63	0.00168

The current density results obtained for these two test series are adequate to achieve optimum contaminant removal, the best values reported by other authors are between 0.0010 to 0.0050 Amp/cm .^{2}

4.4.3 Dissolution of the electrodes

The following equation is used to determine the amount of dissolved iron hydroxide in the electrodes:

$$W = \frac{(D * t * M)}{nF} \qquad (4.1)$$

The results obtained for the first sample group are shown in table 4.47.

Table 4.47 Dissolution of electrodes group 1 (adsorption study).

Sample (ppm)	D (A/cm)^2	W(grs/cm)^2	grs	mgs
1	0.01138889	0.00066091	0.0238	23.8
2	0.01277778	0.00074151	0.0267	26.7
3	0.01361111	0.00078987	0.0284	28.4
4	0.01138889	0.00066091	0.0238	23.8
5	0.01277778	0.00074151	0.0267	26.7
6	0.01388889	0.00080599	0.0290	29
7	0.02083333	0.00120898	0.0435	43.52

The amount of iron hydroxide dissolved from the electrodes is a function of the current density applied to the electrodes, being higher for samples with higher current density applied, which is in agreement with the application of Faraday's law.

For the second set of tests, the electrode dissolution results are shown in table 4.48.

Table 4.48 Dissolution of group 2 electrodes (kinetic study).

Sample (ppm)		$D(A/cm)^2$	$W(grs/cm)^2$	grs	mgs
1	5	0.00173	0.000604	0.226	226
2	10	0.00168	0.000585	0.219	219.36

Similarly, for the second group of samples, the higher the amount of current applied, the higher the dissolution of the electrodes in iron hydroxides.

4.4.4 Energy consumption

The energy consumption of an electrocoagulation reactor can be calculated with equation (2.49). The results obtained are shown in tables 4.50 and 4.51.

Table 4.49 Energy consumption group 1 (adsorption study).

Sample	I (Amp)	V (volt)	t (hr)	E (KWhr)
1	0.41	8.5	0.83	0.000289
2	0.46	12.1	0.83	0.000462
3	0.49	10	0.83	0.000407
4	0.41	12.8	0.83	0.000436
5	0.46	10.7	0.83	0.000408
6	0.50	10.25	0.83	0.000425
7	0.75	6.68	0.83	0.000416

Table 4.50 Energy consumption group 2 (kinetic study).

Sample (ppm)		I (Amp)	V (volt)	t (hr)	E (KWhr)
1	5	0.65	15.43	0.5	0.00591
2	10	0.63	18.75	0.5	0.00501

According to the data obtained from tables 4.49 and 4.50, the energy consumption is another advantage of electrocoagulation compared to the normal techniques used for the removal of arsenic from contaminated water such as electrodialysis or reverse osmosis, these techniques have removal rates higher than 85 %, but their main disadvantage is the high energy consumption which increases the cost of operation.

4.4.5 Cost of treatment

The treatment costs for each electrocoagulation test based on a price of 3.32 pesos per KWhr are shown in tables 4.51 and 4.52.

Table 4.51 Cost of treatment for each sample taken from the test group 1. (Adsorption study).

Sample	Energy cost ($)	E(KWhr)	C_E ($)
1	3.32	0.000289	0.000960
2	3.32	0.000462	0.00153
3	3.32	0.000407	0.00135
4	3.32	0.000436	0.00144
5	3.32	0.000408	0.00135
6	3.32	0.000425	0.00141

7	3.32	0.000416	0.00138

Table 4.52 Cost of treatment for each sample taken from the test group 2. (Kinetic study).

	Sample (ppm)	Energy cost ($)	E(KWhr)	C_E ($)
1	5	3.32	0.00591	0.0196
2	10	3.32	0.00501	0.0166

From tables 4.51 and 4.52 we can see the costs used for each sample, which shows one of the main advantages of electrocoagulation, its low operating cost compared to other chemical or biological systems (the operating cost of the EC process is 60 - 70 % lower than other treatment processes).

4.4.6 Number of moles adsorbed per gram of adsorbent

The number of moles adsorbed per gram of adsorbent was calculated by equation (2.65). The results obtained for sample 1 are shown in table 4.53.

Table 4.53 Number of moles adsorbed.

Sample	C_o (mmol/l)	C (mmol/l)	mc (grs)	N (mmol/g)	C/N
1	0.00725	0.000109	0.0238	0.105	0.00104
2	0.01450	0.000362	0.0267	0.185	0.00196
3	0.03623	0.000507	0.0284	0.440	0.00115
4	0.05072	0.001087	0.0238	0.730	0.00148
5	0.09420	0.002681	0.0267	1.200	0.00223
6	0.1493	0.006884	0.0290	1.666	0.00413
7	0.2174	0.009130	0.0435	1.676	0.00545

The number of adsorbed moles varies according to the concentration of As^{+5} and the amount of dissolved iron hydroxide in the electrodes, this is due to the amount of current applied and the concentration of the samples in each test, the higher the current density applied, the higher the amount of hydroxide to adsorb and the lower the current applied, the lower the amount of dissolved hydroxide. This behaviour can be verified in the following equation (equation to calculate the amount of dissolved iron):

$$W = \frac{(D * t * M)}{nF} \quad (4.2)$$

Where we can observe that the relationship between current density (D) and iron hydroxide generation (W) is directly proportional.

Also, as the concentration of arsenic increases, the probability of adsorption increases. This behaviour can be seen in the results obtained in table 4.53, as the concentration increases, the moles of arsenic adsorbed on the species generated by the electrocoagulation increase:

$$N = V * \frac{Co - C}{m_c} \quad (4.3)$$

In this equation we can verify that the relationship between the concentration of As^{+5} (Co and C) and N (number of moles adsorbed) is directly proportional, which means that as the concentration increases, N increases. The decrease in the % removal of arsenic can also be explained by the fact that the probability of interaction of the

coagulant and cationic ions decreases when they are at low concentrations.
Figure 4.22 shows the Langmuir adsorption isotherm graph,

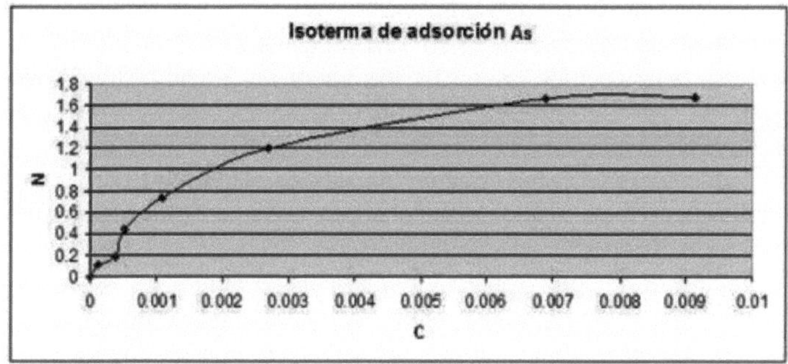

Figure 4.22 Adsorption isotherm of As on iron hydroxides.

C=Final concentration.

N= Number of moles adsorbed.

Using equation (2.67) (linearised Langmuir isotherm equation) and plotting C/N with C (Figure 4.22), we obtain the following linear regression data:

Table 4.54 Data obtained from the linear regression of C/N vs C.

R	0.983
R2	0.966
Ordered	0.0011
Pending	0.4549

If the system follows the behaviour described by the Langmuir isotherm, the graph of the C/N ratio as a function of the equilibrium concentration C should give a straight line of slope 1/Nmax and ordinate to the origin 1/Knmax. The results obtained are:

Table 4.55 Obtaining Nmax and K.

Nmax (mmolAs)/gFe)	2.198
K (Lmmol $)^{-1}$	413.6

The value of Nmax corresponds to the maximum adsorption capacity corresponding to the monolayer fully covered on the surface (mmol/g).

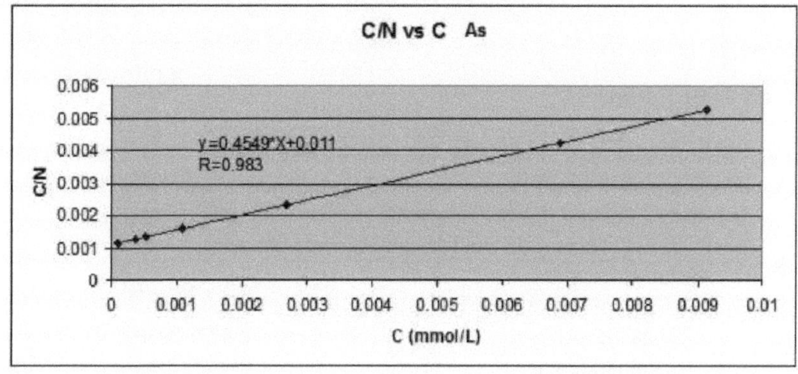

Figure 4.23 Plot of C/N vs. C. The line that best fits the data is shown (C=Final concentration; N=Number of moles adsorbed).

According to the data observed in figure 4.23 we can see that the adsorption system data follow the behaviour described by the Langmuir isotherm (correlation factor 0.983).

4.4.7 Specific area

If the area occupied by each molecule on the surface of the adsorbent is known, it is possible to calculate the specific area of the iron hydroxide (the specific area is the total surface area of one gram of adsorbent).

The BET method of nitrogen adsorption was used to determine the specific area, which assumes that the nitrogen gas when liquefied and adsorbed on solid surfaces will fill all the available clean surface area forming multiple layers. The results obtained for nitrogen adsorption are shown in the table below:

Table 4.56 Nitrogen adsorption data.

Relative Pressure Po/P	Amount adsorbed (cm^3 /g)
0.28442521	53.9894756
0.56090098	60.3178407
0.87575582	67.1412891
1.18879493	73.9187064
1.50239687	80.2187533

In the following figure we have the graph of the BET isotherm corresponding to the nitrogen adsorption.

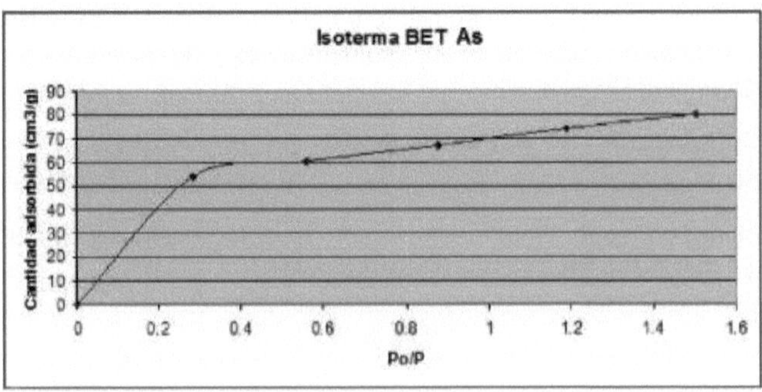

Figure 4.24 BET chart.

To calculate the specific surface area, the data obtained (table 4.58) is plotted using the linearised equation (2.68) BET, then linear regression is applied to the data and using equations (2.72) and (2.73) the specific surface area is calculated.

Table 4.57 Data to calculate specific area.

Relative Pressure Po/P	1/[Q(Po/P - 1)]
0.28442521	0.02792958
0.56090098	0.05237036

0.87575582	0.07906586
1.18879493	0.10549461
1.50239687	0.13386845

Table 4.58 Data obtained from the linear regression.

Pending	0.0865
Ordered to the origin	0.0034
R	0.9999
R2	0.9998

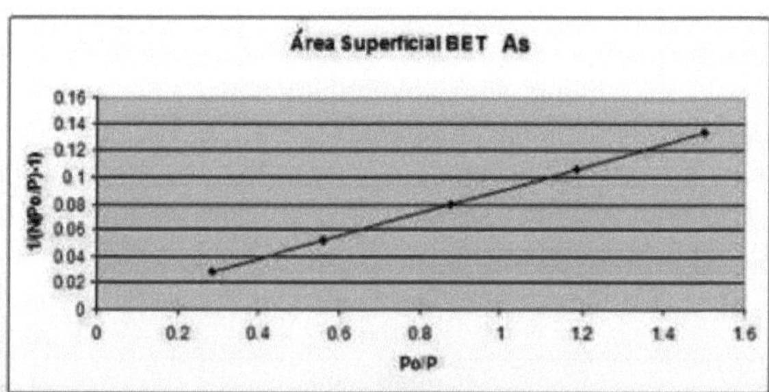

Figure 4.25 Line obtained to calculate the specific area using the BET method.

The value obtained using the BET method is 235 m^2 /g. This value is within the usual range for adsorbents consisting of small porous particles, between 10 and 1 000 m^2 /g (Shoemaker)[58] . The value of the specific area also determines the adsorption capacity of the adsorbent being used, in this case the species generated from EC such as magnetite, goethite, lepidocrocite, etc. This adsorption capacity is checked by the arsenic removal rates obtained.

4.4.8 Fraction of surface area covered

According to the results obtained for N and N_{max}, the fraction of surface area covered Q is calculated using the equation $Q = N/N_{max}$, the results obtained are shown in table 4.59.

Table 4.59 Obtaining Q.

Sample	N	N_{max}	Q
1	0.1049659	2.1983	0.04775519
2	0.18523042	2.1983	0.08427226
3	0.44026842	2.1983	0.2003041
4	0.7299659	2.1983	0.33210459
5	1.19972317	2.1983	0.54582492
6	1.66604198	2.1983	0.75798088
7	1.67566217	2.1983	0.76235767

In table 4.59 it can also be seen that the coating fraction approaches 1 as the concentration increases. The figure below shows the Q data in graphical form.

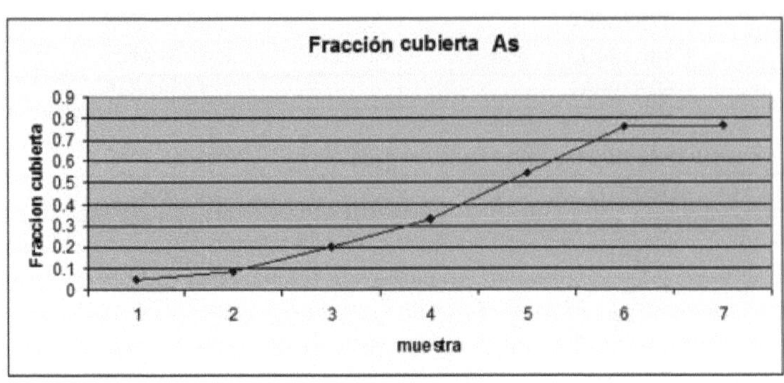

Figure 4.26 Covered fraction Θ.

4.4.9 Calculation of thermodynamic parameters

Applying equations (2.74), (2.75) and (2.76) we obtain the following results for free energy, enta^a and entrop^a for the adsorption process using the 7 concentrations (1,2, 5, 7, 13, 20 and 30 ppm arsenic).

Table 4.60 Thermodynamic parameters.

	Kcal/mol	Kj/mol
AG	-7.63206263	-37.2445
AH	-11.1929	-54.6215
	Kcal/mol K	KJ/mol K
AS	-0.011989	-0.05850

The negative value of AG presented in the table confirms the feasibility of the adsorption process and the spontaneous nature of As adsorption on the species generated from electrocoagulation. The negative value of HA indicates the exothermic nature of the process. The adsorption HA value according to the literature corresponds to typical physisorption values (Typical adsorption enthalpy values for physisorption is -20 kj/mol and about -200 kj/mol for chemisorption[57]), the value obtained being -54.6215 kJ/mol, so the GA value is within the range of monolayer adsorption. The negative AS value reflects that no significant change occurs in the internal structure of the adsorbent during arsenic adsorption.

4.4.10 Calculation of kinetic parameters As

To calculate the kinetic parameters, the second set of tests (tables 4.63 and 4.68) was used, with concentrations of 5 ppm (sample 1) and 10 ppm (sample 2) of As.

4.4.10.1 Calculation of kinetic constants and adsorption constant As 4.4.10.1.1.1 Sample 1

The time-concentration data to determine the kinetic and adsorption constants are shown in the following tables.

Table 4.61 Time-concentration data sample 1.

Time (min)	[As] ppm
0	5
2	3.8
4	2.6
6	1.85

8	0.95
10	0.1

Exponential regression is performed on the data in table 4.61 and the results shown in table 4.62 are obtained.

Table 4.62 Exponential regression data.

R	0.9781
R^2	0.9556
A	5.2283
M	0.195

Applying equation (2.85) the reaction rate is calculated, to calculate the kinetic and adsorption constants the linearised Langmuir-Hinshelwood equation is used, the data for the linear regression are shown in table 4.63.

Table 4.63 Reaction rate data.

-dC/dt	-dt/dC	1/C
0.9815	1.0188487	0.2
0.74594	1.3405904	0.26315789
0.51038	1.95932442	0.38461538
0.363155	2.75364514	0.54054054
0.186485	5.36236158	1.05263158
0.01963	50.942435	10

The data obtained from the linear regression are shown in table 4.64.

Table 4.64 Data obtained from the linear regression.

R	0.9872
R^2	0.9743
a	1.604
M	2.7313

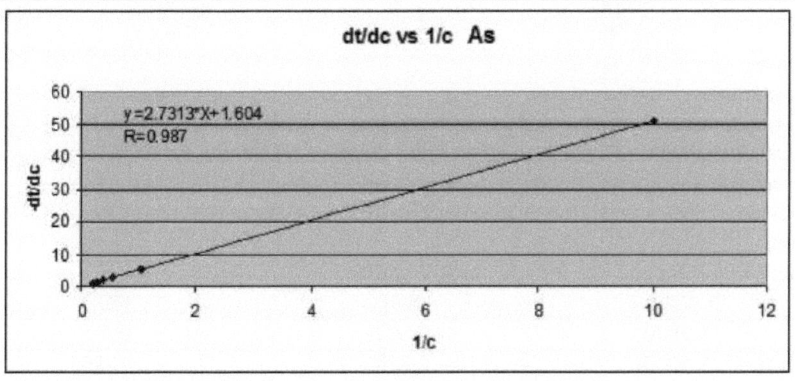

Figure 4.27 Plot of -dt/dc vs 1/c to obtain kinetic and adsorption constants.

and adsorption constants.

Using equations (2.82) and (2.83) the kinetic and adsorption constants are calculated, the results are given in table 4.65.

Table 4.65 Kinetic and adsorption constants.

k (kinetic constant) mgL^{-1} min^{-1}	0.6234414
K (adsorption constant) Lmg^{-1}	0.58726614

For this sample it is observed that the kinetic constant (k) is higher than the adsorption constant (K), which means that the controlling mechanism or the slowest stage for the arsenic-iron hydroxides adsorption process is the physiadsorption or adsorption rate (stage 2 of the adsorption process) rather than the surface reaction rate (stage 3 of the adsorption process).

4.4.10.1.2 Sample 2

The variation of the time-concentration data to determine the kinetic and adsorption constants for the arsenic sample corresponding to a concentration of 10 ppm is shown in the following tables.

Table 4.66 Time-concentration data sample 2.

Time (min)	[As] ppm
0	10
2	7.9
4	5.7
6	3.4
8	1.6
10	0.05

Exponential regression is performed on the data in table 4.66 and the results shown in table 4.67 are obtained.

Table 4.67 Exponential regression data.

R	0.9711
R^2	0.9534
a	10.6381
m	0.1983

Where:

R= correlation coefficient

R^2 = squared correlation coefficient a = ordinate to the origin m= slope.

Applying equation (2.77) the reaction rate (-dc/dt) is calculated, to calculate the kinetic and adsorption constants the linearised Langmuir-Hinshelwood equation, equation (2.78) is used, the data to perform the linear regression are shown in table 4.68.

Table 4.68 Reaction rate data for linear regression.

-dC/dt	-dt/dC	1/C
1.9830	0.5042	0.1
1.5666	0.6383	0.1265
1.1303	0.8847	0.1754
0.6742	1.4831	0.2941

0.3173	3.1517	0.6250
0.00992	100.857	20

The data obtained from the linear regression are shown in table 4.71.

Table 4.69 Data obtained from the linear regression.

R	0.9872
R2	0.9772
a	15.1885
m	5.3485

Where:

R= correlation coefficient

R^2 = squared correlation coefficient a = ordinate to the origin m= slope.

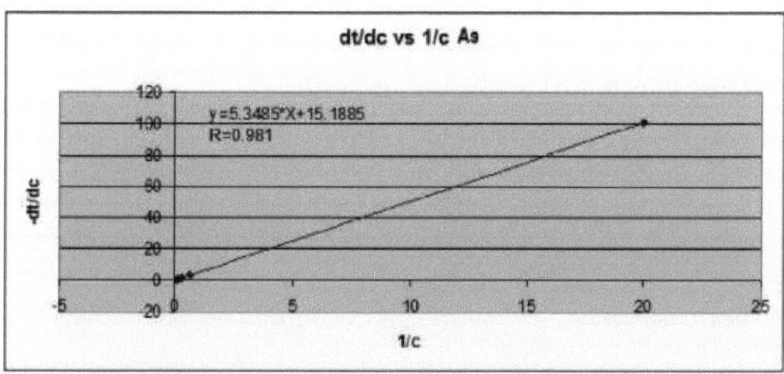

Figure 4.28 Plot of -dt/dc vs 1/c to obtain kinetic and adsorption constants.

and adsorption constants.

Using equations (2.82) and (2.83) the kinetic and adsorption constants are calculated, the results are given in table 4.70.

Table 4.70 Adsorption and kinetic constants.

K (kinetic constant) mgL^{-1} min^{-1}	0.08333333
K (adsorption constant) Lmg^{-1}	2.83976816

For this sample it is observed that the kinetic constant (k) is lower than the adsorption constant (K), which means that the controlling mechanism or slowest stage for the arsenic-iron hydroxides adsorption process is the surface reaction rate (stage 3 of the adsorption process) rather than the physiadsorption or adsorption rate (stage 2 of the adsorption process).

4.4.10.2 Evaluation of the reaction order

The most widely used model to describe the kinetics of the adsorption process is the Langmuir-Hinshenlwood model (Cassano, 1999). To know the order of the reaction, equation (2.77) is used, which is the expression of the first order kinetics, its integrated form is (($-\ln(C_f/C_o)=kt$). When evaluating the kinetic data by means of this model, the following results were obtained.

4.4.10.2.1 Sample 1

To know the order of the reaction, $\ln(C_f/C_o)$ is calculated, where Cf is the final concentration and C_o is the initial concentration, then the obtained data is applied linear regression to know the correlation coefficient and in this way verify the order of the reaction 1.

Table 4.71 Data to determine the reaction order.

Time (min)	Cf (PPm)	Co (ppm)	Cf/Co	-ln(cf/co)
0	5	5	1	0
2	3.8	5	0.76	-0.27443685
4	2.6	5	0.52	-0.65392647
6	1.85	5	0.37	-0.99425227
8	0.95	5	0.19	-1.66073121
10	0.1	5	0.02	-3.91202301

The data obtained from the linear regression are shown in the table below:

Table 4.72 Results obtained from the linear regression.

M	0.4141
A	-0.9855
R	0.976
R2	0.9596

Where:

R= correlation coefficient

R^2 = squared correlation coefficient a = ordinate to the origin m= slope.

The resulting linear regression plot is shown in figure 4.29.

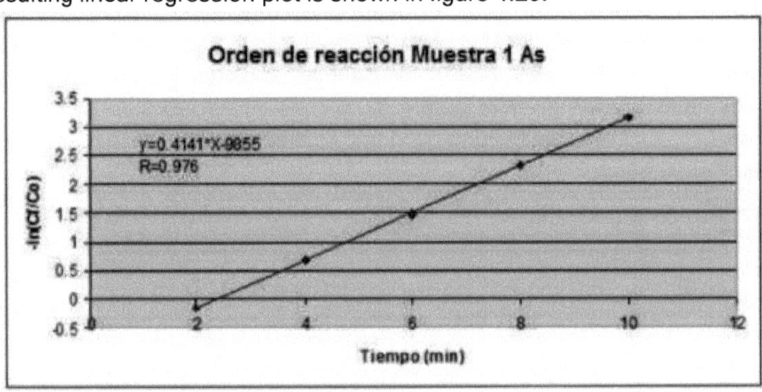

Figure 4.29 -ln(cf/co) vs time plot to determine the order of reaction.

These results verify that the data obtained fit an order 1 model.

4.4.10.2.2 Sample 2

To calculate the order of the reaction, ln(cf/co) is calculated, where Cf is the final concentration and co is the initial concentration, then the data of ln(cf/co) and time are linearly regressed to find the correlation coefficient and thus verify the order of reaction 1.

Table 4.73 Data to determine the reaction order.

Time (min)	Cf (PPm)	Co (ppm)	Cf/Co	-ln(cf/co)

0	10	10	1	0
2	7.9	10	0.79	0.23572233
4	5.7	10	0.57	0.56211892
6	3.4	10	0.34	1.07880966
8	1.6	10	0.16	1.83258146
10	0.05	10	0.005	5.29831737

The data obtained from the linear regression are shown in the table below: Table 4.74 Results obtained from the linear regression.

Pending	0.4543
Ordered to the origin	-0.771
R	0.943
R2	0.889

The resulting linear regression plot is shown in figure 4.30.

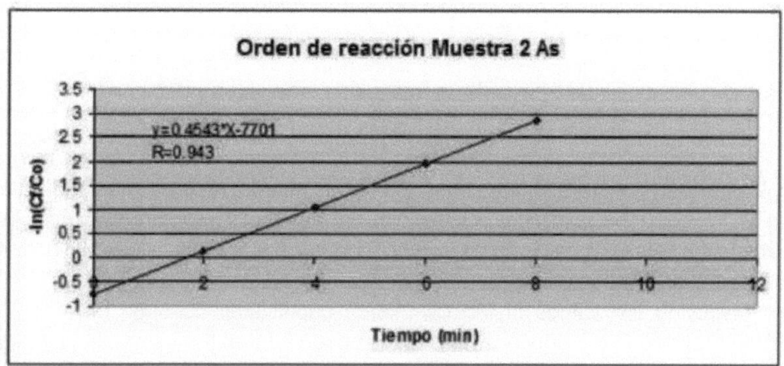

Figure 4.30 -ln(Cf/Co) vs time plot to determine reaction order.

These results verify that the data obtained fit an order 1 model.

4.5 Comparison of adsorption isotherms

To verify that the Langmuir isotherm is the one that best fits the adsorption data we used 2 other isotherms, we compared the correlation factor, the isotherm that best fits the data is the one that has the best correlation coefficient, we also considered another statistical parameter, the average absolute % deviation, the isotherm that has the lowest % deviation is the one that best fits the data The other isotherms used are the Freundlich and D-R isotherms.

(Dubinin-Radushkevich) because after Langmuir they are the most used isotherms in adsorption processes.

4.5.1 Comparison of isotherms using the adsorption data for TiO_2

Table 4.75 Langmuir Isotherm.

Langmuir Isotherm				
R	R 2^A	% Deviation	K (Lmmol $)^{-1}$	Nmax (mmolg $)^{-1}$
0.994	0.988	0.0023	35.12	96

The values shown in table 4.75 are those reported in the previous tables, and are the values that were used to calculate K (Langmuir constant), from this value the thermodynamic parameters were calculated, also the Nmax was calculated, which is the maximum adsorption capacity in this case of titanium dioxide on the particles generated from EC in this case magnetite.

Table 4.76 Freundlich Isotherm.

Freundlich Isotherm				
R	R 2^A	% Deviation	K ($L^{1/n}$ $mmol^{1,1/n}$ $g^{'1}$)	N
0.989	0.978	4.5568	52.0114	1.42

To calculate K from the Freundlich isotherm equation (2.61) was used, the statistical results were obtained using the sigmaplot program, this program includes the exponential function to calculate K and n which is the adsorption intensity and is a parameter that indicates when the adsorption system is favourable when n>1, therefore and according to the obtained value of n (1.42) the titanium dioxide-iron hydroxide adsorption system represents favourable adsorption conditions.

Table 4.77 D-R Isotherm.

D-R isotherm				
R	R 2^A	% Deviation	K (mol kJ $)^{2-2}$	Nmax
0.971	0.942	5.631	3.155	442.611

To calculate the K of the D-R isotherm we first calculated £ (Polanyi potential equal to RT In (1 + 1/Ce)). The statistical assistant of the sigmaplot program was also used, obtaining the values shown, Nmax is the theoretical saturation capacity and K is related to the adsorption energy.

When reviewing the values of the correlation coefficients obtained and the % of deviation reported in the previous tables, it is observed that the Langmuir isotherm is the one that best fits the experimental data, both for the correlation coefficient that is better in the 3 cases and also because it presents the lowest % of deviation of average error, proving with these data the greater use of the Langmuir isotherm for the adsorption calculations and to know the thermodynamic parameters.

4.5.2 Isotherm comparison using As adsorption data

Table 4.78 Langmuir Isotherm.

Langmuir Isotherm				
R	R 2^A	% Deviation	K	Nmax
0.983	0.966	0.0003	413.6	2.198

These values are reported in tables 4.69 and 4.70, they are the results of the linear regression applying the linear form of the Langmuir isotherm, from these data the K is calculated, which was used to know the thermodynamic parameters, also it is calculated from the results of the linear regression the Nmax is the maximum adsorption capacity in this case of arsenic on the generated particles of EC in this case magnetite.

Table 4.79 Freundlich Isotherm.

Freundlich Isotherm				
R	R 2^A	% Deviation	K ($L^{1/n}$ mmol g $)^{1-1/n-1}$	n
0.980	0.960	0.1559	17.33	2.082

The Freundlich equation is an empirical equation used to describe heterogeneous systems, which is characterised by the heterogeneity factor 1/n, (equation 2.61). The value of n represents the adsorption intensity and when n>1 indicates that the adsorption system is favourable, therefore, according to the obtained value (2.082) the arsenic-iron hydroxide adsorption system generated from the EC represents favourable conditions. The statistical parameters were obtained by applying linear regression to the experimental data using the sigmaplot program.

Table 4.80 D-R Isotherm.

D-R isotherm				
R	R 2^A	% Deviation	K	Nmax
0.910	0.828	0.1553	4.70	17.53

The D-R isotherm is more general than the Langmuir isotherm, because it does not assume a homogeneous surface. To know K and Nmax, first the Polanyi potential is calculated, this is plotted with N and the statistical parameters are obtained by applying linear regression.

When comparing the results obtained, we can observe that in both cases, the Freundlich and D-R isotherms have lower correlation coefficients than the Langmuir isotherm, in the case of the % deviation, the isotherm that best fits the data is the one with the lowest %, being in this case also the Langmuir isotherm. With these results it is verified that the Langmuir isotherm is the one that best fits the experimental data.

4.6 Characterisation of the solid product obtained from EC with TiO_2

4.6.1 Characterisation of the solid product obtained from EC with $TiO2$ using R-X diffraction

After filtration, the solid product was dried in an oven at 90 degrees Celsius, then analysed under the scanning electron microscope and X-ray diffraction in order to check for the presence of titanium dioxide and the species generated from electrocoagulation.

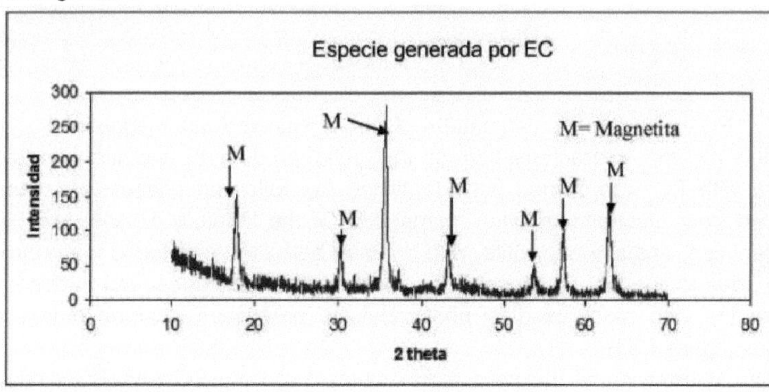

Figure 4.31 Magnetite diffraction pattern generated by EC.

Figure 4.31 shows the diffraction pattern of the solid product obtained from a CD test without any other component in the aqueous medium, only CD species were generated, in order to identify the species generated, this species was identified as magnetite.

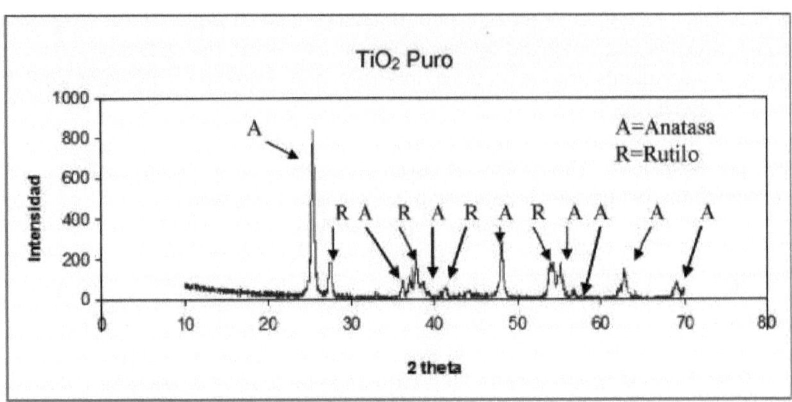

Figure 4.32 Titanium dioxide diffraction pattern.

A diffractogram was then performed on the titanium dioxide used in the cyanide oxidation test only, to identify the phases present, anatase and rutile were identified (figure 4.32).

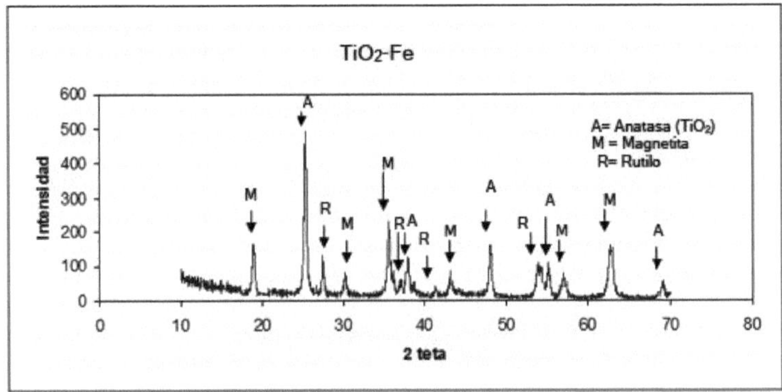

Figure 4.33 Diffraction pattern of titanium dioxide with magnetite.

Subsequently, the characterisation of the solid product of the titanium dioxide recovery with EC was carried out, identifying the following phases: as a species generated from electrocoagulation: magnetite. Of the titanium dioxide, two phases were identified, anatase and rutile, with anatase being the one found in the greatest quantity, due to the fact that most of the peaks identified correspond to this phase, which is the one most used in photocatalysis processes, and rutile in smaller quantities (figure 4.33).

4.6.2 Characterisation of the solid product obtained from EC with TiO2 using the Scanning Electron Microscope

Subsequently, the solid product obtained from the electrocoagulation process was analysed using the Scanning Electron Microscope, the micrographs obtained are shown in the following figures.

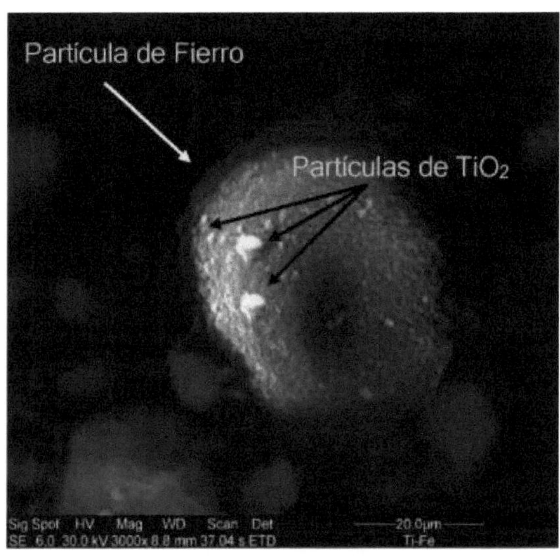

Figure 4.34 Iron particle with TiO2 adsorbed on its surface.

Figures 4.34 and 4.35 show the micrographs obtained with the scanning electron microscopy technique, in which the iron particles generated by EC are observed, these particles are found in greater quantity and size, on these particles are adsorbed the titanium dioxide particles that are going to be recovered for reuse in the photocatalytic oxidation of cyanide.

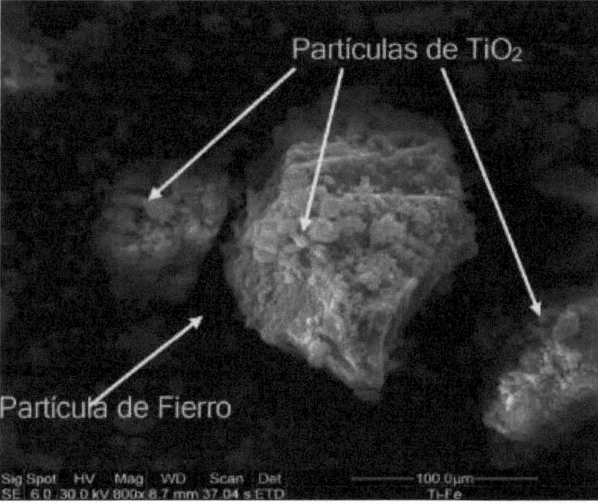

Figure 4.35 Iron particle with TiO2 impregnated on its surface.

Micrographs with this technique show porous particles of electrocoagulation-generated species such as magnetite, lepidocrocite, goethite, impregnated with titanium dioxide.

To verify that the species present are iron in greater quantities and that the titanium dioxide particles are adsorbed on these, chemical analysis by energy dispersive X-ray analysis (EDAX) was carried out (Figure 4.36), thus verifying that the titanium dioxide is found in smaller quantities and is adsorbed on the magnetite particles generated from the CD.

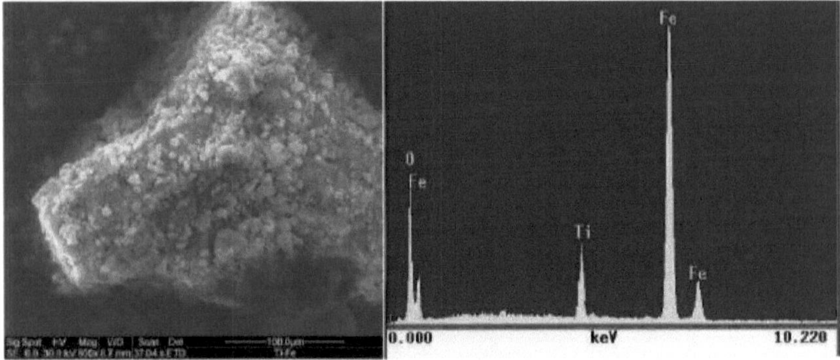

Figure 4.36 Energy dispersive X-ray chemical analysis for Fe-Ti.

4.7 Characterisation of solid product obtained only from CD

Electrocoagulation tests were also performed without any other component of the aqueous medium to generate only electrocoagulation species. Figure 4.37 shows a micrograph of the solid product obtained using the scanning electron microscopy technique; this micrograph was subjected to EDAX elemental chemical analysis (Figure 4.38).

Figure 4.37 Micrograph of an iron particle showing the surface section.

Figure 4.38 EDAX analysis of the iron particle.

Figure 4.37 shows the micrograffa obtained from the iron from the EC, this shows the surface section, on this surface the adsorption of the particles or components of the aqueous medium is carried out, either to recover (titanium dioxide) or to eliminate arsenic, it is also verified by chemical analysis by energy dispersive X-ray that the particle is iron and does not present any other compound (figure 4.38).

4.8 Characterisation of the solid product with Arsenic

4.8.1 Characterisation of the Arsenic-containing solid product using R-X diffraction

After filtration the solid product was dried in an oven at 90 degrees, the powders obtained were analysed under the scanning electron microscope and X-ray diffraction in order to check the presence of the species generated from electrocoagulation.

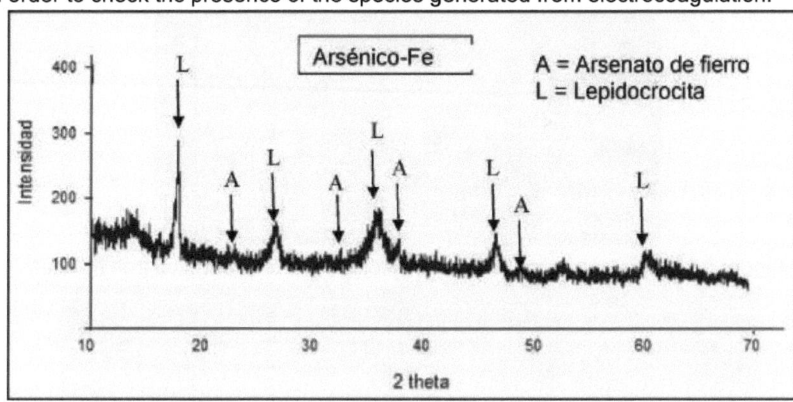

Figure 4.39 X-ray diffraction pattern of the solid product obtained from the treatment of arsenic contaminated water.

Using the X-ray diffraction technique, the species generated from electrocoagulation was identified as lepidocrocite, and it is in this species that the adsorption of the pollutant takes place. The arsenic was identified as iron arsenate. Figure 4.39 shows the X-ray diffraction pattern of the 5 ppm arsenic sample.

4.8.2 Characterisation of the arsenic-containing solid product using the Scanning Electron Microscope

The solid product obtained from the electrocoagulation process for arsenic removal was analysed by scanning electron microscopy, the micrographs obtained are shown in the following figures.

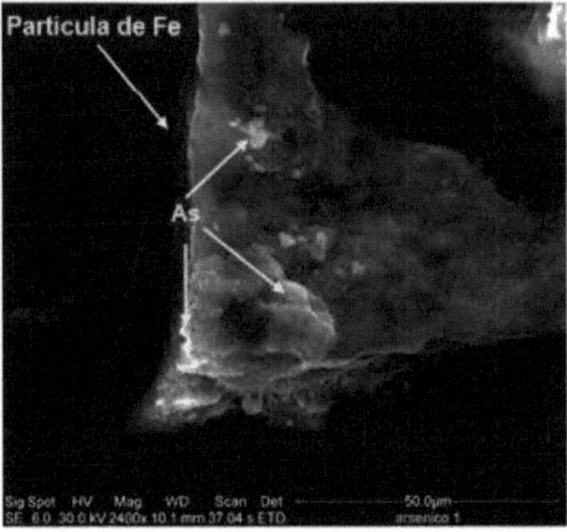

Figure 4.40 SEM micrograph of an arsenic-impregnated iron particle.

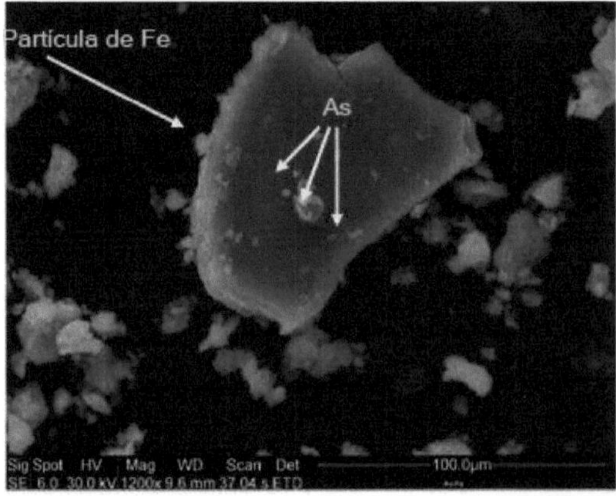

Figure 4.41 SEM micrograph of particles of arsenic-impregnated iron species

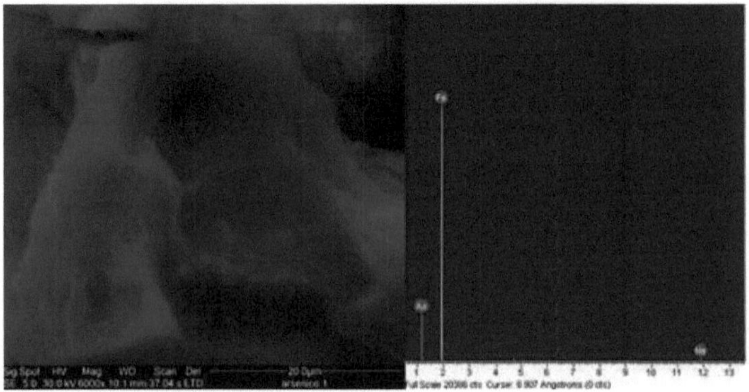

Figure 4.42 Energy Dispersive X-ray Analysis (EDAX).

The micrographs obtained using scanning electron microscopy (figures 4.40 and 4.41), show the porous particles of arsenic-impregnated EC-generated species such as magnetite, goethite and goethite. Figure 4.42 was chemically analysed by energy dispersive X-ray analysis (EDAX) to check the presence of arsenic in the iron particles. By means of this analysis it is verified through the higher count of the analysis that most of the particle is iron and in smaller quantity arsenic is found, which is adsorbed in the iron species generated by electrocoagulation.

V CONCLUSIONS

1. With the photocatalytic oxidation process, using titanium dioxide as oxidising semiconductor, 94 % cyanide elimination was obtained.
2. With the electrochemical process of Electrocoagulation, a 98% recovery of titanium dioxide from the cyanide solution was obtained.
3. According to these results, it is proved that the electrocoagulation technology is an economically viable option to recover titanium dioxide nanocrystals from the photocatalytic oxidation of cyanide.
4. The thermodynamic analysis of the parameters obtained for the recovery of titanium dioxide are as follows: AG= -29.0184 KJ/mol, AH=-13.397 KJ/Mol and AS=0.0524 KJ/molK. The negative value of the free energy indicates the spontaneous nature of the adsorption. The value of the adsorption enthalpy corresponds to the values of a physiadsorption mechanism (enthalpy values of -20 KJ/mol).
5. The Langmuir thermodynamic model perfectly models the results to predict the adsorption capacity of titanium dioxide on iron species generated by electrocoagulation. In this model the following adsorption data were determined: the number of adsorbed moles (N value from 13.605 to 77.966 mmolTio2/gFe), maximum adsorption capacity (Nmax=96.7 mmolTio2/gFe), specific area (190 m2/g) and the fraction covered *Q* (0.139 to 0.812).
6. For the samples used (concentrations of 0.5, 0.75 and 1 g/L titanium dioxide), the kinetic study was carried out by varying the concentration with time, the kinetic constant (k) is higher than the adsorption constant (K). This means that the controlling phenomenon for the TiO2 recovery process with electrocoagulation is the rate of adsorption of titanium dioxide (stage 2 of the adsorption process).
7. The order of the reaction for titanium dioxide recovery followed the Langmuir-Hinshelwood first order kinetic model.
8. The characterisation of the CD product using X-ray diffraction technique, was identified as an iron species known as magnetite and the titanium dioxide was identified as Anatase and Rutile. The latter is found in smaller quantities.
9. With the scanning electron microscope, the presence of titanium dioxide on the surface of the iron particles was verified using the energy dispersive X-ray analysis (EDAX) technique.
10. In the electrocoagulation tests carried out at laboratory level to remove arsenic, a removal efficiency of 99 % was obtained.
11. These results prove that electrocoagulation is a viable and economical alternative to remove arsenic from contaminated water.
12. The Langmuir model was applied with good results to predict the adsorption capacity of arsenic on iron species, the following adsorption calculations were made: the number of adsorbed moles with N values from 0.103 to 1.676 mmolAs/gFe, maximum adsorption capacity, Nmax=2.198 mmolAs/gFe, spatial area of 235 m2/g and the fraction covered *3* from 0.048 to 0.760.
13. The thermodynamic parameters, obtained for arsenic removal, are as follows: AG = -37.245 KJ/mol and AH = -54.6215 KJ/mol. The negative value of AG indicates the spontaneous nature of the adsorption process. The value of AH corresponds to the physiadsorption process.

14. For the kinetic study, the sample with a concentration of 5 ppm of As was used, in this case the kinetic constant (k) is higher than the adsorption constant (K), therefore the controlling step is the adsorption rate (step 2 of the adsorption process).
15. For the second sample, a concentration of 10 ppm As was used, in this case the kinetic constant (k) is lower than the adsorption constant (K), therefore the controlling step is the surface reaction rate (step 3 of the adsorption process).
16. The order of the reaction for arsenic removal followed the Langmuir-Hinshelwood first order kinetic model.
17. From the X-ray diffraction characterisation, different electrocoagulation species were identified as goethite, magemite, lepidocrocite, arsenic was identified as hydrated hydrogen arsenate and iron arsenate.
18. With the scanning electron microscope, the presence of arsenic on the surface of the iron hydroxide particles was checked by energy dispersive X-ray analysis (EDAX).
19. With the comparison of isotherms (Langmuir, Freundlich and D-R), by means of the statistical parameters of correlation coefficient and % deviation, it was verified that the isotherm that best fits the data is the Langmuir isotherm.

VI SOURCES OF INFORMATION

1 http://www.plata.com.mx/Plata/Plata/produccionintem.htm (Accessed January 2007).
2 M. I. Jeffrey, I. M. Ritchie, The Leaching of Gold in Cyanide Solution in the Presence of Impurities, J.Electrochem.Soc. I. The Effect of Lead. 147 (9) 3257-3262. (2000).
3 J.R. Parga and D.L. Cocke, Oxidation of Cyanide in The Hydrocyclone Reactor, Journal of Desalination, 140 289-296. (2001).
4 J.D. Desai, C. Ramakrishna, P.S. Patel, S.K Awasthl, Cyanide Wastewater Treatment and Commercial Applications, Chemical Engineering World. XXXIII (6) 115-121. (1998).
5 G.H. Robbins, Historical Development of the INCO SO_2/air Cyanide Destruction Process, CIM Bulletin (9) 63-69. (1996).
6 J.R. Parga, D.L. Cocke, J.L. Valenzuela, J.A. Gomes, M. Kesmez. Arsenic Removal via Electrocoagulation from Heavy Metal contaminated groundwater in La Comarca Lagunera, Mexico. Journal of Hazardous Materials B124. 247-254. (2005).
7 Greenberg, Arnold et al. Standard methods for the examination of water and wastewater. Washington D.C.: American public health association, 268 p. ISBN 0-87553-131-8. (1985).
8 G. Pavas E. Camargo M. P., Castro Jones C, Pineda V.T, Oxidacion FotocataKtica de Cianuro, ISSN 1692-0694, Paper 29-042005, Universidad EAFIT, MedelKn, Colombia, April (2005).
9 V. Augugliaro, J. Blanco, J. Caceres, E. Garcia, V. Loddo, Photocatalytic oxidation of Cyanide in Aqueous TiO2 Suspensions Irradiated by Sunlight in Mild and Strong Oxidant Conditions, Catalysis Today, 54 245-253. (1999).
10 U.B. Ogutueren, E. Toru, S. Koparal, Removal of Cyanide by Anodic Oxidation for Wastewater Treatment, Water Research. 33 (8) 1851-1856. (1999)
11 11 Young, C.A. and Jordan, T.S. Cyanide remediation: current and past technologies: Proceedings of the 10th annual conference on hazardous waste research. Vol. 44, 104-129. (1999). 12 U.B. Ogutueren, E. Toru, S. Koparal, Removal of Cyanide by Anodic Oxidation for Wastewater Treatment, Water Research. 33 (8) 1851-1856. (1999).
12 13 W.L. Perry, W.W. Eckenfelder, Toxicity Reduction in Industrial Effluents. Eckenfelder, Toxicity Reduction in Industrial Effluents (New York: Van Nostrand Reinhold Co. (1990). 14 Norma Oficial Mexicana NOM - 003 - ECOL - 1997. 15 G.H. Robbins, Historical Development of the INCO SO_2/air Cyanide Destruction Process, CIM Bulletin (9) 63-69. (1996). 16 Longsdon Mark, Kagelstein Karen, I. Mudder Terry. The management of cyanide in gold mining. International Council on Metals and the Environment. ICME (Chem.129 National Council on Metals and Environment), (2003).
13 17 J. A. Dirk Van Zyl, P. G Hutchison, Introduction to Evaluation, Design, and Operation of Precious Metal Heap Leaching Projects, Society of Mining Engineers of AIME, Technology & Industrial Arts, 298-303. (1998) 18 Legrini, E. Oliveros and A.M. Braun, Chem. Rev., 93, 671-698,(1993). 19 Huang CP, Ch. Dong and Z. Tang, Waste Management, 13, 361-377, (1993). 20 US/EPA Handbook of Advanced Photochemical Oxidation Processes, EPA/625/R-98/004, (1998).
14 21 The AOT Handbook, Calgon Carbon Oxidation Technologies, Ontario, (1997).

15 22 Bolton J.R. and Cater S.R., "Aquatic and Surface Photochemistry", 467-490.
16 G.R. Helz, R.G. Zepp and D.G. Crosby Editors. Lewis, Boca Raton, Florida, USA, (1994).
17 23 Glaze W.H., Environ. Sci. Technol., 21, 224-230,(1987). 24 Glaze W.H.,. Kang J.W and Chapin D.H., Ozone Sci. & Technol., 9, 335-352, (1987).
18 25 González Silvia, Sahores Martha. limpacto ambiental debido al uso de cianuro en la minería a cielo abierto. National University of Patagonia, San Juan Bosco, Faculty of Natural Sciences (2003). 26 G. Pavas E. Camargo M. P, Castro Jones C, Pineda V.T, Oxidación Fotocatalítica de Cianuro, ISSN 1692-0694, Document 29-042005, Universidad EAFIT, Medellín, Colombia, April (2005). 27 Ahumada Theoduloz Gerardo. CI 51 K drinking water treatment processes, Apuntes de coagulación-floculación, Universidad de Chile, (2005).
19 http://cipres.cec.uchile.cl/~ci51k/Apuntes/CoagulacionFloculacion.pdf (Date of
20 accessed March 2007)
21 28 Sharma V.K. Rivera W., Joshi VN., Millero FJ. Environ. Sci. Technol., 33, 2645-2650, (2000).
22 29 Fujishima, A. ; Rao, T. N. ; Tryk, D. A. Titanium dioxide photocatalysis. J.
23 Photochem. & Photobio. C: Photochem. Rev., 1, 1-21, (2000). 30 V. Augugliaro, J. Blanco, J. Caceres, E. Garcia, V. Loddo, Photocatalytic oxidation of Cyanide in Aqueous TiO_2 Suspensions Irradiated by Sunlight in Mild and Strong Oxidant Conditions, Catalysis Today, 54 249-253. (1999) 31 D. Bhakta, S. S. Shukla, M.S. Chandrasekharalah, J. L. Margrave, A Novel Photocatalytic Method For Detoxification of Cyanide Wastes, Environ. Sci.
24 Techno. 26 625-626. (1992). 32 U.B. Ogutueren, E. Toru, S. Koparal, Removal of Cyanide by Anodic Oxidation for Wastewater Treatment, Water Research. 33 (7) 1850-1854. (1999).
25 33 P. Liang, Y. Qin, B. Hu, CH. Lin, Tianyou Peng, Z. Jiang, Study of the Behavior of Heavy Metal Ions on Nanometer-Size Titanium Dioxide With ICP-AES, Fresenius J. Anal. Chem. 368 638-640. (2000). 34 R. Shapiro, S. Dubelman, A.M Feinberg, Heterogeneous Photocatalytic Oxidation of Cyanide Ion in Aqueous Solutions at TiO_2 Powder, Departmen of Chemistry, New York University, 303-304. (1990).
26 35 G. Pavas E. Camargo M. P., Castro Jones C, Pineda V.T, Photocatalytic Oxidation of Cyanide, ISSN 1692-0694, Paper 29-042005, Universidad EAFIT, Medellin, Colombia, April (2005). 36 Blanco J. Malato, Bahnemann D. Carmona F. Y Martinez F. Proceedings of 7th Inter.. symp. On Solar Thermal Conc. Tech., IVTAN Ed. ISBN 5-201- 09540-2, 540-550, Moscow,Russia, (1994). 37 N. A. Jaramillo, Photodegradation, SENA La Salada, Caldas Antioquia, 4,7.
27 38 D. Dabrowsky, J. Hupka, M. Zurawzka, Laboratory and Pilot Scale Photodegradation of Cyanide-containing wastewater, Phisicochemical Problems of Mineral Processing, 39, 229-248. (2005)
28 39 http://www.miliarium.com/monografias/arsenico/ (Accessed October
29 2007)
30 40 Scott J.P.y. Ollis D.F, Environ. Progress, 14, 88-103,(1995). 41 Deng B., Burris DR. and Campbell TJ, Environ. Sci. Technol., 33, 2651- 2556, (2000).
31 42 http ://www.ingenieroambiental.com/informes/arsenicoestudio.htm (Date from

32 accessed May 2007). 43 Kovatcheva, V.K. and Parlapanski, M.D. Sono-Electrocoagulation of Iron Hydroxides, Colloids and Surfaces, 149, 603-608. (1999). 44 Mollah, M., Schennach R., Parga J.R. and Cocke D. L., Electrocoagulation (EC)-Science and Applications, Journal of Hazardous Materials, B84 29-41. (2001).

33 45 http://www.ingenieroambiental.com/informes/arsenicoestudio2.htm (Date of

34 accessed February 2008) 46 Vlyssides, A.G., Papaioannou, D., Loizidoy, M., Karlis. P.K., and Zorpas, A.A. Testing an Electrochemical Method For Treatment of Textile Dye Wastewater, Waste Management, 20, 569-574. 47 Mills D. A New Process For Electrocoagulation, American Water Works Association, 92, 6, 39-45. (2000).

35 48 G. Chen, X. Chen, P. L. Yue, Electrocoagulation and Electroflotation of Restaurant Wastewater, J. Environmental Engineering, Sept. 858-862.

36 49 M. J. Matteson, R. L. Dobson, R. W. Glenn, N. S. Kukunoor, Electrocoagulation and Separation of Aqueous Suspensions of Ultrafine Particles, J. colloids and Surfaces, 104 101-109. (1995). 50 Kovatcheva, V.K. and Parlapanski, M.D. Sono-Electrocoagulation of Iron Hydroxides, Colloids and Surfaces, 149, 603-608. (1999). 51 Parga, J.R., Shukla, S,S. and Carrillo-Pedroza, F.R. Destruction of Cyanide Waste Solution Using Chlorine Dioxide, Ozone and Titania Sol. (2003). 52 H.A. Moreno, D.L. Cocke, A.G Gomez, P Morkovsky, J.R. Parga.

37 Electrocoagulation Mechanism for COD Removal. Separation and Purification Technology. SEPPUR-D-06. 11-14. (2007) 53 J.R parga,D.L. Cocke,V Valverde,J.AG. Gomes,H Moreno, and D. Mencer.

38 Characterization of Electrocoagulation for Removal of Cr and As. Journal of 132hem.... Eng Technol. 28. 605-612. (2005). 54 M.Y.A. Mollah,P. Morkosvy, A.G. Gomes, M. Kesmez. D.L. Cocke.

39 Fundamentals, Present and Future Perspectives of Electrocoagulation. J.

40 Hazard Mater. B114. 199-210. (2004) 55 Legrini, E. Oliveros and A.M. Braun, Chem. Rev., 93, 671-698,(1993). 56 G. Vicuña I. Tuñon. Apuntes de Química Avanzada, Department of Chemistry-Physics. University of Valencia. March (2006). 57 R. Ramos, L.G Velazquez, R.M. Guerrero. Adsorption of sodium salicylate in aqueous solution on activated carbon. Journal of the Mexican Chemical Society. Numero 002, Vol. 46. 159-166. (2003). 58 I. Tuvert, V. Talanquer. On adsorption. Para Saber Experimentar y Simular , Educación Química, Facultad de Quimica, UNAM, 6. 186-190. (1999).

41 59 D. F. Shoemaker. Experiments in Physical Chemistry, Mc-Graw Hill (1998) 60 A.G Gupta, S Kundu. Adsorptive Removal of As(III) from Aqueous Solution Using Iron Oxide Coated Cement (IOCC): Evaluation of Kinetic Equilibrium and Thermodynamic Models. Separation and Purification Technology, 51, 165-172 (2006).

42 61 G. Pavas E. Camargo M. P., Castro Jones C, Pineda V.T, Oxidación Fotocatalítica de Cianuro, ISSN 1692-0694, Document 29-042005, Universidad EAFIT, Medellín, Colombia, April (2005). 62 G. Vicuña I. Tuñon. Apuntes de Química Avanzada, Department of Chemistry-Physics. University of Valencia. March (2006). 63 Elías Chávez J. A. Metallurgical comparison of cyanide regeneration using aluminium. Master's thesis in materials. Instituto Tecnológico de Saltillo, February (2000).

Printed by Books on Demand GmbH, Norderstedt / Germany